STRENGTH OF MATERIALS

BOOKS BY JOHN W. BRENEMAN:

Mathematics, Second Edition
Mechanics, Third Edition
Strength of Materials, Third Edition

STRENGTH OF MATERIALS

John W. Breneman, C.E., P.E.

PROFESSOR OF ENGINEERING MECHANICS

Prepared under the Direction of the Division of Continuing Education
THE PENNSYLVANIA STATE UNIVERSITY

Third Edition

McGRAW-HILL BOOK COMPANY
New York · St. Louis · San Francisco · London
Toronto · Sydney · Mexico · Panama

STRENGTH OF MATERIALS

Printed in the United States of America.

Library of Congress Catalog Card Number: 64-66200

07–007536–0

11 12 13 14 15 KPKP 7 9 8 7

FOREWORD

This volume is one of three texts prepared by the author, a member of the College of Engineering staff of The Pennsylvania State University. These books are of value in college classes, but their principal purpose is for use in adult education and engineering associate degree programs.

Experience in technical education indicates that there is a lack of teaching materials suitable for instruction in this field. Such classes at The Pennsylvania State University are composed of individuals who wish to study material which has fairly direct application. This is particularly true of students working in industry who study part-time, as well as those who are preparing for jobs in industry. The schooling of many of these students is mostly self-education and, in many cases, an experience in industry which constitutes a valuable apprenticeship. These texts have also proved to be invaluable for those whose formal training beyond high school is geared to terminal programs of shorter duration than four-year university curriculums.

Technical education instructors are selected on the basis of their practical experience in a particular field as well as academic preparation. It follows, therefore, that text materials used in such classes should be readable, understandable, and practical.

The author of *Strength of Materials*, J. W. Breneman, Professor of Engineering Mechanics, College of Engineering, has incorporated in this volume some of the results of a wide experience in the fields of industrial engineering training as well as in teaching at the college level. While some theory is included, emphasis is placed on the applications of principles to important practical problems.

E. L. KELLER
Director of Continuing Education
The Pennsylvania State University

PREFACE

In this, the third edition of *Strength of Materials*, the author has endeavored to amplify the student's understanding of the basic principles of the subject. In so doing, new and different approaches to the limited theory have been introduced. In several cases, additional theory has been added to clarify the fundamental idea being developed.

Greater emphasis has been placed on the deformations which accompany the stresses, in the hope that the importance of strains becomes better understood as a limiting condition of design. Wherever suitable, the general concept of strain energy is included to emphasize the fact that energy absorption and release are important concepts of the elastic behavior of materials.

The author has not overlooked the basic purpose of the text; namely, that it is designed and written for the technician and engineering aide. Mathematics beyond the level of most high schools is not required.

Further, the author believes in acquainting the students with theory by applying it. The applications are made by solving problems where calculations are not the major emphasis. In many problems, the answers can be completed without the use of long and tedious calculations. All solutions should be completed, for it is by a careful examination and interpretation of the answer that one acquires a feel for the practical magnitude of the terminal result.

I made use of my many years of teaching and practical experience in formulating the many problems which appear at the end of each chapter. A list of answers to some of the problems is given at the back of the text. This helps the student to understand when he is applying the correct theory and to acquire some degree of confidence in what he has learned. The problems without answers are necessary if the student is to realize that he alone must learn to produce a correct result.

I should like to express my sincere thanks to the many persons who have contributed suggestions and helpful ideas, and particularly the American Institute of Steel Construction for their permission to reproduce the tables of properties of WF sections. My sincere appreciation is due Miss Kathryn Nelson for the laborious hours spent in typing and in checking the details of the manuscript for this text.

My everlasting gratitude is due my very understanding wife for her infinite patience while she was a "book widow."

JOHN W. BRENEMAN

CONTENTS

1 STRESS AND STRAIN 1

1-1 Definitions, Stress, Strain
1-2 Stresses
1-3 Loads
1-4 Static Loads
1-5 Stresses
1-6 Factor of Safety
1-7 Strain
1-8 Stress-Strain Behavior
1-9 Poisson's Ratio
1-10 Biaxial and Triaxial Strain
1-11 Strain Energy
1-12 Triaxial Strain Energy
1-13 Temperature Effects

2 SHEAR AND CONNECTIONS 31

2-1 Shear Stress
2-2 Riveted Joints
2-3 Riveted Joint Failures
2-4 Shear
2-5 Tension
2-6 Bearing
2-7 Efficiency of Riveted Joints
2-8 Pressure Vessels
2-9 Structural Riveting
2-10 Welded Joints

3 TORSION 61

3-1 Torque
3-2 Shearing Stresses
3-3 Angle of Twist in Shafts
3-4 Strain Energy
3-5 Power Transmission by Shafts
3-6 Shaft Couplings

4 BEAMS—SHEAR AND MOMENT DIAGRAMS 81

4-1 Beams
4-2 Beam Supports
4-3 Types of Beams
4-4 Loads on Beams
4-5 Beam Reactions
4-6 Shear at a Section of a Beam
4-7 Moments and Moment Diagrams
4-8 Résumé

5 STRESSES IN BEAMS 103

5-1 Bending Stresses
5-2 Design of Beams
5-3 Shearing Stress in Beams
5-4 Horizontal Shearing Stress
5-5 Shearing Stresses in I Beams

6 BEAM DEFLECTIONS 119

6-1 Bending
6-2 Radius of Curvature
6-3 Beam Deflections
6-4 Bending Strain Energy

7 COMBINED AXIAL AND BENDING STRESSES—ECCENTRIC LOADS ON RIVETED CONNECTIONS 134

7-1 Axial and Bending Stresses
7-2 Bending When the Load Is Not in a Principal Plane
7-3 Biaxially Loaded Riveted Connections
7-4 Eccentrically Loaded Riveted Connections

8 COLUMNS 151

8-1 Columns
8-2 Steel Columns
8-3 Structural Aluminum Columns
8-4 Timber Columns

APPENDIX

Properties of WF Sections 168
Properties of American Standard Channels 173
Properties of Equal Angles 174
Properties of Unequal Angles 176
Elements of Sections 178
Weights and Areas of Bars 181
Screw Threads and Bolts 183
Coefficients of Expansion 185

ANSWERS TO SELECTED PROBLEMS 186

Index 189

STRENGTH OF MATERIALS

CHAPTER 1
STRESS AND STRAIN

1-1. Definitions. Strength of materials is that branch of the science of engineering which studies the effects of forces acting on pieces of deformable materials and the resulting deformations. Deformable materials are used for parts of machines and engineering structures.

Stress is the internal resistance of a material to external force(s). When the external force(s) exceed the internal resisting stresses, the material will rupture or break.

Strain or deformation is the change in the dimensions or configuration of a deformable body.

1-2. Stresses. There are three types of fundamental stresses, known as *tension*, *compression*, and *shearing stresses*. These can occur in a particle of material singly or in various combinations, and one type can cause another type of stress to exist. Thus, what sometimes appears to be a very elementary condition of stress analysis or design may require a more thorough analysis to provide a safe condition of the material.

Tension, or a tensile stress, is caused by a force which tends to pull out or stretch a piece. The cables on a crane are subjected to a tensile stress when a load is being lifted.

Compression, or a compressive stress, is caused by a force tending to push together or shorten a piece. The base of a heavy machine is in compression, for the entire weight of the machine bears down and tends to crush the base.

Shear, or a shearing stress, is caused by forces which tend to cause one part of a piece to slide over another part. Punching holes in boiler plate, or cutting any sheet metal, is an example of shear.

Figures 1-1 to 1-3 show how the forces act in these simple stresses. In Fig. 1-1, the forces tend to pull the piece apart so that it is in tension. In Fig. 1-2, the forces tend to crush the piece so that it is in compression. In Fig. 1-3, some of the forces tend to move the metal one way, while the other forces tend to move it in the opposite direction. The two parts of the piece, therefore, tend to slide over each other, or to shear at the middle, as shown by the dotted line. This action, at Fig. 1-3, is just the same as if a pair of shears were cutting the material.

In Figs. 1-1 and 1-2, the collinear forces which are acting on the bodies are perpendicular, or normal, to the cross-sectional area. They are applied so that the line of action of the force coincides with the geometrical axis of the body and are known as *axial forces.* In Fig. 1-3, the forces are applied parallel but noncollinear to the cross-sectional area, causing the shearing action.

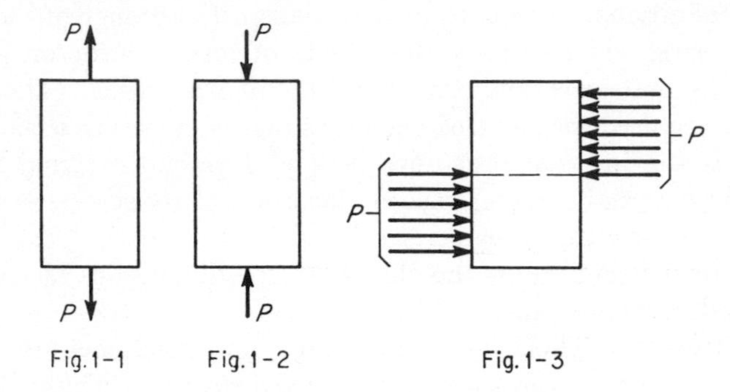

Fig. 1-1 Fig. 1-2 Fig. 1-3

1-3. Loads. The external forces which set up stresses in a body are called *loads.* The load on a bridge, therefore, will be the weight of its own members plus the weight of anything passing over it. The most common form of load which acts upon a piece of material is the *static,* or *dead, load,* and it is the principal type of loading that will be considered in this text. The effect of *impact loads,* which are suddenly applied forces, is included in the subject matter when it can be appropriately introduced.

A type of loading which must be considered in a more comprehensive text is *repeated loads,* which alternately increase and decrease and introduce fatigue stresses that may lead to fatigue failure. These alternations can vary from a few to millions of cycles of repetition.

1-4. Static Loads. Forces which are applied gradually to the piece of material and remain essentially constant after being applied are called static loads. The weight of a building on the foundations will be a static load in compression. Wires stretching between poles have a static load in tension. The rivets in a tank will have a static load in shear if the pressure in the tank always remains the same. Figure 1-4 shows why a rivet will always have a shearing action. The pressure in the cylinder tends to flatten out the plate so that the pull on the rivet is as shown by the arrows. This pull tends to shear the rivet off between the plates. (See Chap. 2 for a more extended discussion of the analysis of the stresses in riveted connections.)

1-5. Stresses. The intensity of a force normal to any cross-sectional area is called the *normal stress* at a point. Figure 1-5 shows a weightless piece of material subjected to equal, opposite, and collinear axial forces P and P. The piece or member is in a state of static

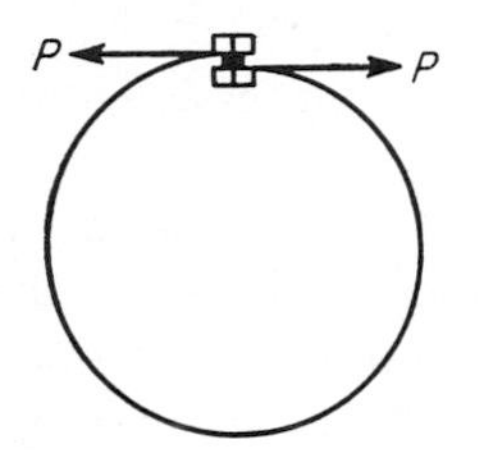

Fig. 1-4

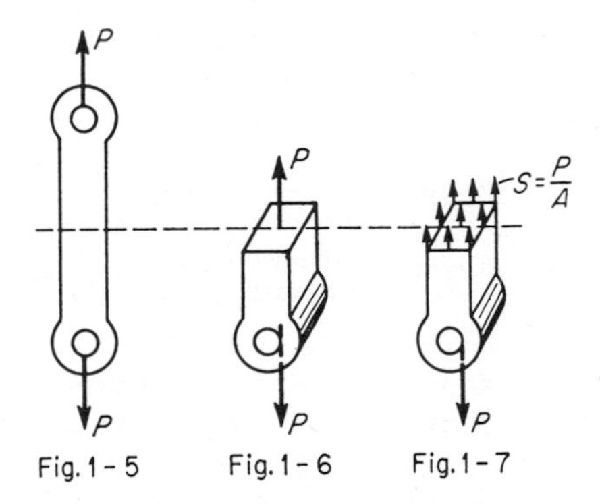

Fig. 1-5 Fig. 1-6 Fig. 1-7

equilibrium. Using the principle of a free-body diagram which permits a part of a member subjected to external forces to be isolated for study, Fig. 1-6 shows the lower part of the member also in a state of equilibrium. Then the normal stress is defined by the mathematical equation

$$\text{Force} = \text{total resisting stress}$$

$$P = SA$$

Since forces will always be expressed in pound units and the areas in square inches, it follows from

$$S = \frac{P}{A}$$

that S will be in pounds per square inch (abbreviated psi). The term *stress* will be used solely to express the unit-force condition of pounds per square inch; it is illustrated in Fig. 1-7.

Using Fig. 1-5 for reference, with $P = 36{,}000$ lb acting on the normal cross-sectional area $A = 3$ sq in., the stress

$$S = \frac{P}{A} = \frac{36{,}000}{3} = 12{,}000 \text{ psi}$$

Similar reasoning applies to compressive forces and their corresponding stresses.

The *ultimate strength* of a material is the maximum stress which the material can resist just before it breaks. The ultimate strength may be affected by the rate of application of the loading; for a high rate of loading, the ultimate strength tends to be greater than for a low rate of loading. The values given in Table 1-1 are for low rates of loading.

TABLE 1-1. PROPERTIES OF MATERIALS
Average values in psi

Material	Tensile strength, ultimate	Compressive strength, ultimate	Shear strength, ultimate	Modulus of elasticity	
				Tension and compression	Shear
Wrought iron	50,000		40,000	27,000,000	11,000,000
Structural steel	70,000		45,000	†30,000,000	12,000,000
0.46% carbon steel:					
Hot-rolled	87,000		55,000	†30,000,000	12,000,000
Annealed	70,000		45,000	†30,000,000	12,000,000
Oil-quenched	220,000			†30,000,000	12,000,000
Oil-quenched and tempered	112,000			†30,000,000	12,000,000
Cold-rolled	160,000			†30,000,000	12,000,000
Chrome-nickel steel	100,000		90,000	†30,000,000	12,000,000
Monel metal	85,000		50,000	26,000,000	10,000,000
Gray cast iron		80,000		12,000,000	
Brass	50,000			12,000,000	
Aluminum alloy 2024T	68,000		41,000	10,000,000	4,000,000
Yellow pine		8,500*	1,500*	1,600,000	
Fir		6,000*	1,000*	1,200,000	

* Parallel to the grain.

† Often used as 29,000,000.

1-6. Factor of Safety. In designing a part for a machine or for a structure, the total amount of the load on the piece is first calculated. The piece is then made large enough so that it will withstand this total without being subjected to a stress which will bring about a permanent deformation. *Hence, the allowable stress will always be much less than the ultimate strength of the piece.* If the stress on the piece is only one-half the ultimate strength, it may be said that the piece is only doubly safe; or if the stress is only one-third the ultimate strength, the piece is three times safe. *This relation of the actual or allowable stress in the piece and the ultimate strength of it is called the factor of safety or*

$$\text{Factor of safety} = \frac{\text{ultimate stress}}{\text{actual stress}}$$

Hence, if a chain has an ultimate strength of 50,000 psi and it is to carry a maximum stress of 10,000 psi, then it is five times safe, or its *factor of safety* is 5.

When an allowable stress is desired and the ultimate strength is known, the factor of safety is generally specified by a design code or should be estimated.

Table 1-1 shows the average values of the ultimate strengths and moduli of elasticities of various materials. When additional special values are required, they will be given as they apply. When a value is omitted from the table, it means that the material is generally not used for that kind of stress or that the limiting condition is other than the ultimate stress.

The factor of safety which should be applied will be stated with the example or problem. In some cases, where allowable or design stresses have been recommended as good practice, these will be specified where they apply. Technical groups in their special fields have been instrumental in advancing design procedures for their materials and have brought about considerable standardization of design specifications. The American Society for Testing Materials, The American Institute of Steel Construction, and The National Lumber Manufacturers Association are just a few of the many groups that have been active in the various fields in setting the standards for the proper uses of their materials.

Table 1-1 should be referred to for the values of the strengths of the materials for the several conditions of loading. The steels are used for compressive loading; but the allowable stress is limited by the elastic, or yield, strength, which approximates 0.6 of the tensile ultimate strength. The allowable compressive strengths are considered to be nearly equal to the allowable tensile strengths.

Example. What tensile load would cause a bar of structural steel 2.5 in. × 1.5 in. to fail?

Solution. The cross section of the bar is

$$A = 2.5 \times 1.5 = 3.75 \text{ sq in.}$$

From Table 1-1, structural steel has an ultimate strength of 70,000 psi. Hence, the maximum load will be

$$P = AS$$

$$P = 3.75 \times 70{,}000 = 262{,}500 \text{ lb} \qquad \textit{Ans.}$$

The specified allowable tensile strength for this steel is 20,000 psi. What safe tensile load will this bar support?

$$P = AS$$

$$P = 3.75 \times 20{,}000 = 75{,}000 \text{ lb} \qquad \textit{Ans.}$$

Example. A wrought-iron bar ½ in. in diameter withstands a load of 3,000 lb in compression. Calculate the unit stress.

Solution. The cross-sectional area of a ½-in. bar will be

$$\frac{\pi(\frac{1}{2})^2}{4} = 0.196 \text{ sq in.}$$

Hence $$\frac{3{,}000}{0.196} = 15{,}000 \text{ psi} \qquad \textit{Ans.}$$

In the same manner, the size of a member can be determined if the load is known. For example, suppose a piece of structural steel must be designed so that it will hold a load of 50,000 lb in tension.

From Table 1-1 we find that structural steel has an ultimate tensile strength of 70,000 psi. Assuming that a factor of safety of 3.5 is specified, the allowable stress is

$$\frac{70{,}000}{3.5} = 20{,}000 \text{ psi}$$

Then, to determine the required area,

$$P = AS$$

from which $A = \frac{P}{S}$

$$A = \frac{50{,}000}{20{,}000} = 2.5 \text{ sq in.}$$

If a square bar is desired, it will now be necessary to find the side of a square which has 2.5 sq in. for the area. This is done by extracting the square root of the area, or

$$\sqrt{2.5} = 1.58 \text{ in.}$$

Hence, if the bar is square, it will be 1.58 in. on each side.

If the bar is to be circular instead of square, then we must find the diameter of a circle whose area is 2.5 sq in.

$$\text{Area of circle} = \frac{\pi D^2}{4}$$

Hence

$$2.5 = \frac{3.1416D^2}{4}$$

from which

$$D^2 = 3.18 \qquad \text{or} \qquad D = 1.78 \text{ in.} \qquad \textit{Ans.}$$

The following examples have been worked out so that the student may get a clear idea about the design of pieces subjected to axial forces. These examples should be carefully studied and thoroughly understood before proceeding with the study of the text.

Example. The wrought-iron hoisting chain in a crane has a cross-sectional area of ⅞ sq in. Is it safe for this crane to lift a load of 6 tons?

Solution. The chain will be in tension, and Table 1-1 shows that the ultimate strength of wrought iron in tension is 50,000 psi. The load on a crane is variable, so that the factor of safety should be 5. Hence, the allowable stress will be

$$\frac{50{,}000}{5} = 10{,}000 \text{ psi}$$

Now, since the cross-sectional area of the chain is ⅞ sq in., the allowable load on the chain will be

$$P = AS = \tfrac{7}{8} \times 10{,}000 = 8{,}750 \text{ lb}$$

Therefore, the maximum load this crane should lift is 8,750 lb, so that it will not be safe for it to lift 6 tons, which is 12,000 lb.

Example. A steel tension member on a bridge must be made to withstand a pull of 5,000 lb. Find the diameter of the rod, assuming a factor of safety of 5.

Solution. The allowable stress will be the ultimate strength divided by the factor of safety, or

$$\frac{70{,}000}{5} = 14{,}000 \text{ psi}$$

Hence

$$\frac{5{,}000}{14{,}000} = 0.36 \text{ sq in. required}$$

from which

$$0.36 = \frac{\pi D^2}{4} \qquad \text{or} \qquad D = 0.68 \text{ or about } \frac{11}{16} \text{ in.} \qquad \textit{Ans.}$$

Example. What must be the size of a square cast-iron block to support a weight of 60 tons static-compressive load?

Solution. From Table 1-1 the ultimate strength is found to be 80,000 psi, and a factor of safety of 5 will be used. Hence

$$\frac{80{,}000}{5} = 16{,}000 \text{ psi allowable stress}$$

The total load is 60 tons, or $60 \times 2{,}000 = 120{,}000$ lb. Hence

$$A = \frac{P}{S} = \frac{120{,}000}{16{,}000} = 7.5 \text{ sq in. required}$$

Since the block is square, it follows that

$$\sqrt{7.5} = 2.74 \text{ or } 2\tfrac{3}{4} \text{ in.} \qquad \textit{Ans.}$$

Example. A threaded bolt is shown in Fig. 1-8. When the bolt is tightened by screwing up the nut, a tensile force is applied which slightly stretches the bolt. If a sufficient tensile load P is applied to the bolt, it may break, as shown in Fig. 1-9, across the area of a circle whose diameter is the net (root) diameter of the bolt. Thus, a ¾-in.-diameter bolt, as shown in Fig. 1-9, is stretched by tightening the nut until the bolt breaks, when the total tensile force is 18,700 lb. What is the ultimate unit tensile stress of the material?

From the Appendix, the net diameter c of a ¾-in. bolt is 0.620 in. and the net area (root area) is 0.302 sq in. Then

$$S = \frac{P}{A} \quad \text{and} \quad S = \frac{18{,}700}{0.302} = 61{,}920 \text{ psi}$$

Note. Failure of the bolt may occur by a stripping of the threads, rather than by a rupture of the root area. This failure will be given consideration in Chap. 2.

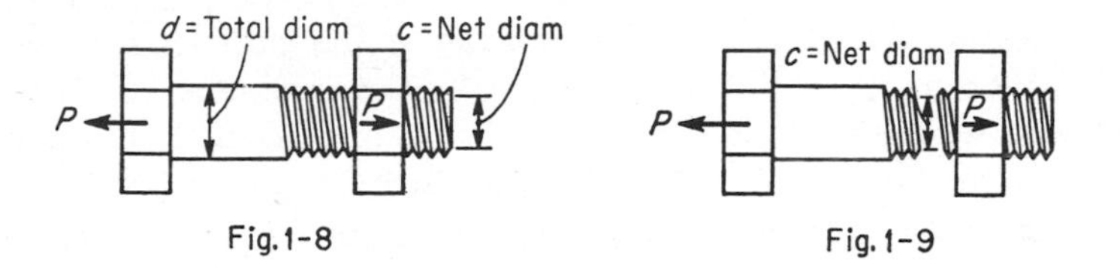

Fig. 1-8 Fig. 1-9

Example. The piston in an engine is 10 in. in diameter, and the steam pressure is 100 psi. Find the proper diameter of the chrome-nickel steel piston rod.

Solution. The area of the piston will be

$$\text{Area} = \frac{\pi D^2}{4} = \frac{3.14 \times 10 \times 10}{4} = 78.54 \text{ sq in.}$$

and the net area of the piston on the side which causes tension in the rod is $(78.54 - A)$ sq in. where A represents the area of the rod.

Since the pressure is 100 psi, the total pressure will be $(78.54 - A)100$ lb.

From Table 1-1, we find that the ultimate strength is 100,000 psi, and we shall use a factor of safety of 10, since the load is applied suddenly. Hence, the allowable stress will be

$$\frac{100{,}000}{10} = 10{,}000 \text{ psi}$$

Now, if the total load on the piston rod is $(78.54 - A)100$ lb and the allowable stress is 10,000 psi, then the area must be

$$A = \frac{P}{S} = \frac{(78.54 - A)100}{10{,}000}$$

from which $$A = \frac{7{,}854}{10{,}100} = 0.7776 \text{ sq in.} = \frac{\pi D^2}{4}$$

and $$D = 0.995 \text{ in.} \qquad Ans.$$

1-7. Strain. In Art. 1-1, it was stated that when a body was subjected to a force, the body was deformed. This deformation is related to the stress which causes it.

It is from laboratory tests that the relations between force and deformation are determined, the most common type of test used being the tensile test. In this test, a piece of material is subjected to a gradually increasing force, and simultaneously the *change* in a measured initial gage length is noted. The change in the gage length is called the total strain and will be designated by the symbol e. Just as it is preferable to think of and use force in its unit form S, it is likewise preferable to use deformation in its unit form.

Denoting the unit deformation by the letter ε (the Greek letter epsilon) and the original length by L, then

$$\varepsilon = \frac{e}{L}$$

To be consistent with the units of stress, which are pounds per square inch, the units of ε will be inches per inch (abbreviated in. per in.). This requires that the units of both e and L be expressed in inches. Thus the deformation per unit length is termed *strain.* For most materials, ε is a very small quantity.

1-8. Stress-Strain Behavior. Structural steel is a ductile material which stretches considerably before fracture. Its stress-strain relationship is best shown by a stress-strain diagram. A typical example of this is shown in Fig. 1-10. The abscissa scale represents strain, and the ordinate values represent the stress scale. A critical examination of the diagram provides certain important and significant values for the designer.

From the origin to point A on the curve, the ratio of stress to strain may be assumed to be constant. This assumption is sufficiently accurate for most materials and the ratio is an expression known as Hooke's law. In symbol form, the ratio

$$\frac{S}{\varepsilon} = \text{a constant} = E$$

where E represents the value called the *modulus of elasticity.* E has the same psi units as stress S.

Point A, which is the limit of the linear relationship of stress to strain, is called the *proportional limit.* The proportional limit corresponds closely to the *elastic limit,* which is defined as the greatest stress a material can develop without a permanent deformation remaining upon complete release of the load. This is interpreted as that stress where permanent deformation begins.

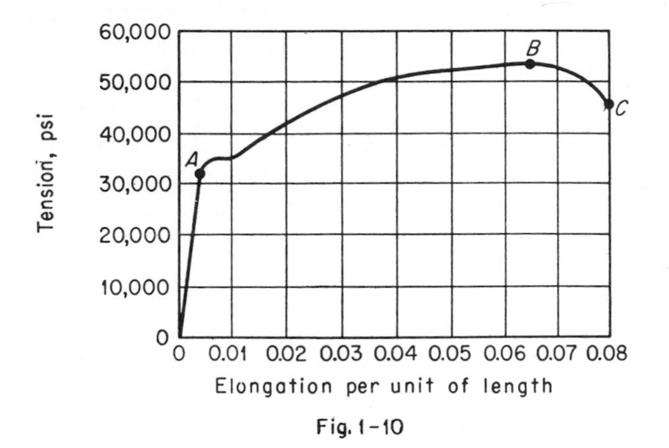

Fig. 1-10

An examination of Table 1-2 will be helpful in understanding the elastic range of behavior. It should be noted that for the last increase of 2,500 lb, the change in length was double that of each of the preceding changes. Thus the proportional (or elastic limit) would be 25,000 lb.

TABLE 1-2

P, lb	e, in.	P, lb	e, in.
0	0	15,000	0.48
2,500	0.08	17,500	0.56
5,000	0.16	20,000	0.64
7,500	0.24	22,500	0.72
10,000	0.32	25,000	0.80
12,500	0.40	27,500	0.96

Again referring to Fig. 1-10, note that in general the part of the curve which extends from the origin to point A is referred to as the *elastic* range of behavior of the material, and the part which extends from point B to point C is known as the *plastic* range of behavior. Of importance in the plastic range are the two points B and C. Point B designates the maximum or *ultimate* stress, while point C indicates the fracture or *breaking* stress.

A piece of cast iron tested in compression will have the same ultimate and breaking loads, since, when the maximum load is reached, the piece will break.

The following example illustrates Hooke's law. A tie rod is made of structural steel for which the modulus of elasticity is 30,000,000 psi. The rod is 12 ft long and is 1 in. × 2 in. in cross section. When a load of 36,000 lb is applied to the rod, how much does it stretch?

From Hooke's law,

$$E = \frac{S}{\varepsilon} \quad \text{but} \quad S = \frac{P}{A} \quad \text{and} \quad \varepsilon = \frac{e}{L}$$

Then

$$E = \frac{P/A}{e/L} = \frac{PL}{Ae}$$

where P = total load, lb
L = *original length*, in.
A = *original area*, sq in.
e = total deformation (elongation or contraction), in.
E = modulus of elasticity, psi

This equation is important as it shows the relation between the applied load and the total deformation. From $E = PL/Ae$,

$$e = \frac{PL}{AE}$$

Then, substituting the given values,

$$e = \frac{36{,}000 \times 144}{2 \times 30{,}000{,}000} = 0.0864 \text{ in. total stretch}$$

and the unit stretch (stretch per inch of length) is

$$\varepsilon = \frac{e}{L} = \frac{0.0864}{144} = 0.0006 \text{ in./in.}$$

1-9. Poisson's Ratio. When an axial force P is applied to pieces of material, it is known that the body is changed in length. With the volume being constant, it becomes apparent that the lateral or transverse (at right angles to the axial load) dimensions must change. When the length is increased, the lateral dimensions will be decreased, and vice versa.

It has been determined experimentally that, for any given material and an elastic condition of stress, there is a constant ratio of lateral strain ε_L to axial strain ε. This ratio is known as Poisson's ratio and is designated by the symbol μ (the Greek letter mu). In formula form, this becomes

$$\mu = \frac{\varepsilon_L}{\varepsilon}$$

Figure 1-11 shows an exaggerated condition of the lateral deformation of a block subjected to axial compressive force. In order to simplify the notation used on the diagram, the deformations are indicated as

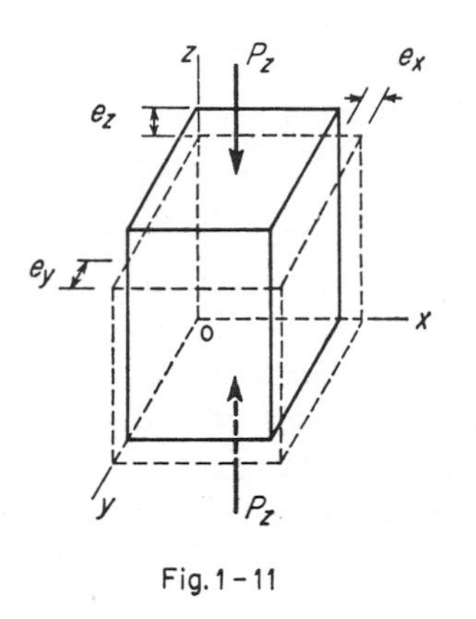

Fig. 1-11

though one corner (o) of the block is fixed in its original position. The change in the L_z dimension is e_z, the change in L_x is e_x, and the change in L_y is e_y. From the expression for strain

$$\varepsilon = \frac{e}{L} \qquad \text{then} \qquad \varepsilon_x = \frac{e_x}{L_x} \qquad \text{and} \qquad \varepsilon_y = \frac{e_y}{L_y}$$

and

$$\mu = \frac{\varepsilon_x}{\varepsilon} = \frac{\varepsilon_y}{\varepsilon}$$

or in general

$$\mu = \frac{\varepsilon_L}{\varepsilon}$$

The numerical values of μ vary from approximately 0.25 to 0.35 with the value of Poisson's ratio for steel being commonly used as 0.30.

Example. What is the change in the 3-in. diameter of a 12-in.-long bar of steel subjected to an axial force of 70.7 tons?

$$\text{Area} = \frac{\pi 3^2}{4} = 7.07 \text{ sq in.} \qquad \text{and} \qquad S = \frac{141{,}400}{7.07} = 20{,}000 \text{ psi}$$

Then
$$\varepsilon = \frac{S}{E} = \frac{20{,}000}{30 \times 10^6} = \frac{2}{3{,}000} \text{ in./in.}$$

Using $\mu = 0.30$, and ε_L for strain in the diametrical direction,

$$\mu = \frac{\varepsilon_L}{\varepsilon} \qquad \text{and} \qquad \varepsilon_L = 0.30 \times \frac{2}{3{,}000} = \frac{0.6}{3{,}000}$$

The change in the diameter is

$$e_L = \varepsilon_L L = \frac{0.6}{3{,}000} \times 3 = 0.0006 \text{ in.}$$

1-10. Biaxial and Triaxial Strain. There are innumerable examples of materials which are simultaneously subjected to stresses in the direction of more than one coordinate axis. An illustration of this is given in Fig. 1-12. The diagram has been simplified by assuming that the material is fixed at one corner (o) with all resulting deformations taking place away from or toward that corner.

The block of material is subjected to tensile stress S_x in the X-axis direction, tensile stress S_y in the Y-axis direction, and compressive stress S_z in the Z-axis direction. This is known as a triaxial stress condition and produces a system of internal stresses, the study and analysis of which are beyond the scope of this text. It will suffice to understand that such an internal system of stresses does exist and that the magnitudes of several of the internal stresses may exceed the magnitudes of the S_x, S_y, and S_z stresses which are applied to the block.

However, just as it is possible to analyze the deformations in each of the coordinate directions resulting from a uniaxial stress, the analysis of the strains in the coordinate directions becomes one of superposition.

The *principle of superposition* implies that the total effect is equivalent to the sum of the effects from the separate causes. Applying this to the elastic strains resulting from the applications of S_x, S_y, and S_z, then

$$\text{Total } \varepsilon_x = \pm \varepsilon_x \pm \mu\varepsilon_y \pm \mu\varepsilon_z$$

with similar equations being available for the total strains in the Y and Z directions. The plus or minus signs are used to indicate that either sign may apply, depending on the increase or decrease in size of the block. It is customary to use a plus sign for an increase and a minus sign for a decrease.

For the tensile S_x shown, ε_x is plus in the X direction with $-\mu\varepsilon_x$ in the Y and Z directions. For the tensile S_y, ε_y is plus in the Y direction with $-\mu\varepsilon_y$ in the X and Z directions. For the compressive S_z, ε_z is

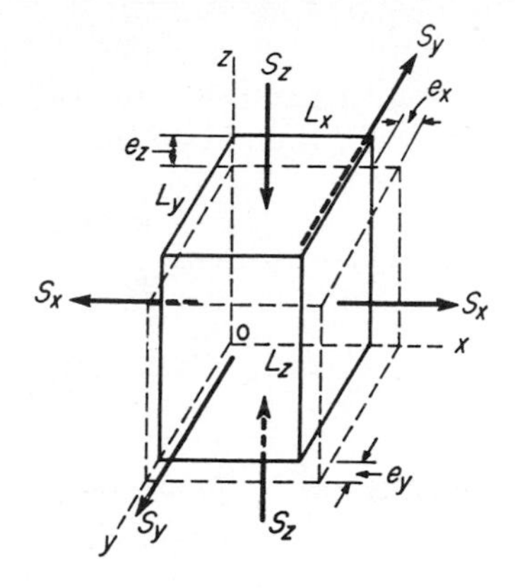

Fig. 1-12

minus in the Z direction with $+\mu\varepsilon_z$ in the X and Y directions. In equation form, these statements become

$$\text{Total } \varepsilon_x = +\varepsilon_x - \mu\varepsilon_y + \mu\varepsilon_z$$

$$\text{Total } \varepsilon_y = +\varepsilon_y - \mu\varepsilon_x + \mu\varepsilon_z$$

$$\text{Total } \varepsilon_z = -\varepsilon_z - \mu\varepsilon_x - \mu\varepsilon_y$$

Recalling that the modulus of elasticity E equals the ratio of stress to strain, then

$$\varepsilon = \frac{S}{E}$$

Substituting this value of ε in the three equations above, they become

$$\text{Total } \varepsilon_x = \frac{S_x}{E} - \mu\frac{S_y}{E} + \mu\frac{S_z}{E}$$

$$\text{Total } \varepsilon_y = \frac{S_y}{E} - \mu\frac{S_x}{E} + \mu\frac{S_z}{E}$$

$$\text{Total } \varepsilon_z = -\frac{S_z}{E} - \mu\frac{S_x}{E} - \mu\frac{S_y}{E}$$

and

$$e_x = (\text{total } \varepsilon_x)\, L_y$$

$$e_y = (\text{total } \varepsilon_y)\, L_y$$

$$e_z = (\text{total } \varepsilon_z)\, L_z$$

It must be realized that elastic strains are usually very small, and the total changes in length likewise small.

When the stress in the direction of any one of the coordinate axes is zero, the corresponding resulting unit strains will be equal to zero.

For the condition of $S_y = 0$, the above three *total* unit-strain equations reduce to

$$\text{Total } \varepsilon_x = \frac{S_x}{E} + \mu \frac{S_z}{E}$$

$$\text{Total } \varepsilon_y = -\mu \frac{S_x}{E} + \mu \frac{S_z}{E}$$

$$\text{Total } \varepsilon_z = -\frac{S_z}{E} - \mu \frac{S_x}{E}$$

A numerical example will serve to illustrate this by an application of the preceding theory. Assume that the following data apply to Fig. 1-12.

Let

$L_x = 2$ in.; $L_y = 3$ in.; $L_z = 4$ in.

$S_x = S_y = +16{,}000$ psi tension

$S_z = -12{,}000$ psi compression (note the use of the conventional signs with the stresses)

$E = 30 \times 10^6$ psi $\qquad \mu = 0.30$

What is the change in each of the coordinate lengths?

$$\text{Total } \varepsilon_x = +\frac{16{,}000}{30 \times 10^6} - 0.30\frac{16{,}000}{30 \times 10^6} + 0.30\frac{12{,}000}{30 \times 10^6}$$

$$\text{Total } \varepsilon_y = +\frac{16{,}000}{30 \times 10^6} - 0.30\frac{16{,}000}{30 \times 10^6} + 0.30\frac{12{,}000}{30 \times 10^6}$$

$$\text{Total } \varepsilon_z = -\frac{12{,}000}{30 \times 10^6} - 0.30\frac{16{,}000}{30 \times 10^6} - 0.30\frac{16{,}000}{30 \times 10^6}$$

from which

$\text{Total } \varepsilon_x = +0.000493$ in./in. and $e_x = +0.000986$ in.

$\text{Total } \varepsilon_y = +0.000493$ in./in. and $e_y = +0.00148$ in.

$\text{Total } \varepsilon_z = -0.00072$ in./in. and $e_z = -0.00288$ in.

1-11. Strain Energy. So far, the emphasis on the design of engineering parts has been dependent on the application of static loads. In many cases, the parts are required to absorb the energies due to suddenly applied forces. An example of an energy load is the effect of a moving particle of mass which is suddenly stopped. This effect is generally termed *impact.* When the mass is suddenly stopped, all or part of its energy is transferred to the object it strikes, and the resulting combination of stress and strain in the object is called *strain energy.*

It should be recalled from the study of the principles of work that the work done by a lineally increasing (from zero) force is equal to the product of the *average* force and the displacement in the direction of

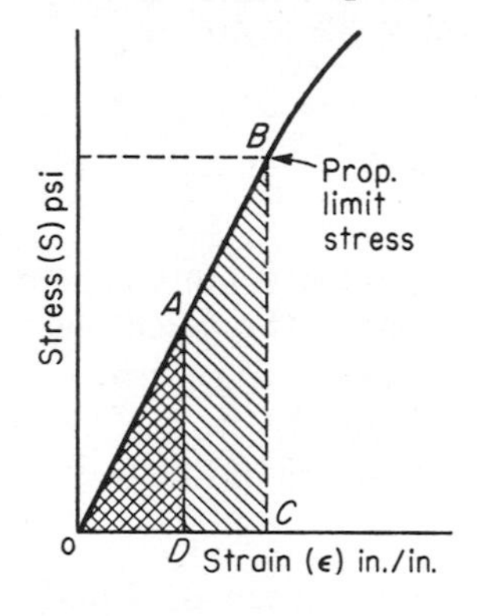

Fig. 1-13

the force. When the senses of both the force and displacement are the same, the work is positive. This is the exact relation which exists between an axial force and the corresponding axial deformation. Letting U_t represent the total work, then the total work in inch-pound units is

$$U_t = \tfrac{1}{2}Pe$$

From

$$P = SA \qquad \text{and} \qquad e = \varepsilon L$$

$$U_t = \tfrac{1}{2}SA \cdot \varepsilon L = \tfrac{1}{2}S\varepsilon \cdot AL$$

Since AL is equivalent to the volume, in cubic inches, of the piece of material under an axial static load, and $\tfrac{1}{2}S\varepsilon$ is in inch-pounds per cubic inch, U_t will be expressed in the correct units of inch-pounds.

Strain energy is customarily thought of as a unit condition or as inch-pounds per cubic inch. Representing unit strain energy by U,

$$U = \frac{U_t}{\text{volume}} = \tfrac{1}{2}S\varepsilon = \frac{1}{2}\frac{S^2}{E}$$

In Fig. 1-13, when S equals the stress at the proportional limit, the unit strain energy equals the shaded area under the *elastic* part of the stress-strain diagram. This is an important concept of recoverable energy from an elastically strained material.

As an example of the energy required to strain a certain part to its elastic limit, assume that a ¾-in.-diameter steel bar 10 in. long is to be stressed to its 36,000-psi elastic limit. What is the total energy required, *or* if this stress were released, what is the total recoverable elastic energy?

$$U = \frac{1}{2}\frac{S^2}{E} = \frac{(36{,}000)^2}{2 \times 30 \times 10^6} = 21.6 \text{ in.-lb/cu in.}$$

and $$U_t = U \times \text{volume} = 21.6 \times 0.442 \times 10$$

$$U_t = 95.5 \text{ in.-lb}$$

Just as it is essential to apply a factor of safety to stresses in order to ensure safe operating parts, the same reasoning should be applied when designing parts to resist energy forces. As an example of this, assume that the part referred to in the preceding example is to be designed to absorb only one-third of its elastic-energy capacity. Then

$$U_t = \frac{95.5}{3} = 31.8 \text{ in.-lb}$$

and $$U = \frac{U_t}{\text{volume}} = \frac{31.8}{4.42} = \frac{1}{2}\frac{S^2}{E}$$

Then $$S^2 = \frac{31.8 \times 2 \times 30 \times 10^6}{4.42} = 43.2 \times 10^7$$

and $$S = 20{,}800 \text{ psi}$$

which corresponds approximately to point A on the curve. Note that this stress does not equal the 36,000-psi stress divided by the factor of safety. Rather, it equals the stress at A such that the double-shaded area (OAD) is one-third of the single-shaded triangle (OBC).

If the factor of safety of 3 is applied to the stress, then

$$U_t = \frac{1}{2}\frac{S^2}{E} \times \text{volume} = \frac{(12{,}000)^2}{2 \times 30 \times 10^6} \times 4.42$$

$$= 10.6 \text{ in.-lb}$$

1-12. Triaxial Strain Energy. It logically follows that if axial strain energy can be expressed as

$$U = \tfrac{1}{2} S \varepsilon$$

then the triaxial strain energy will be equivalent to the sum of the elastic strain energies created by the stresses and strains in the directions of the three coordinate axes. Thus

$$U = \tfrac{1}{2} S_x \varepsilon_x + \tfrac{1}{2} S_y \varepsilon_y + \tfrac{1}{2} S_z \varepsilon_z$$

where ε_x, ε_y, and ε_z represent the totals of the unit strains in the X, Y, and Z directions. Figure 1-14 represents a piece of material subjected to $+S_x$, $+S_y$, and $+S_z$ stresses (all tensile). This system of

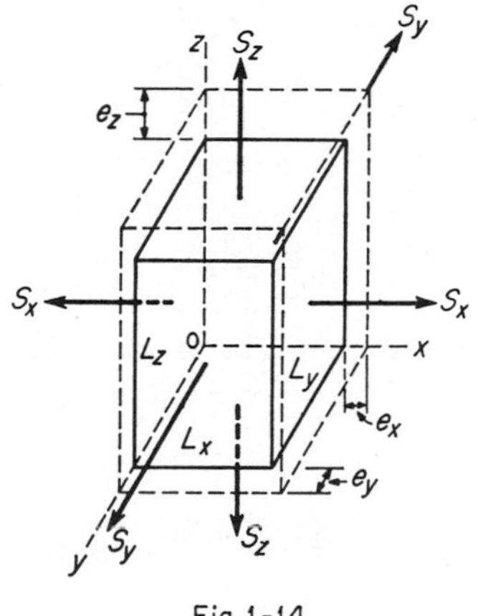

Fig. 1-14

stresses is used to obtain a general strain-energy condition which is applicable to any combination of stresses. Then any stress can be zero, or any stress can be compressive by merely using a minus value in the equation.

$$\text{Total } \varepsilon_x = \frac{S_x}{E} - \mu \frac{S_y}{E} - \mu \frac{S_z}{E}$$

$$\text{Total } \varepsilon_y = \frac{S_y}{E} - \mu \frac{S_x}{E} - \mu \frac{S_z}{E}$$

$$\text{Total } \varepsilon_z = \frac{S_z}{E} - \mu \frac{S_x}{E} - \mu \frac{S_y}{E}$$

Combining the four preceding equations, the strain energy (inch-pounds per cubic inch)

$$U = \frac{1}{2E}\left[S_x(S_x - \mu S_y - \mu S_z) + S_y(S_y - \mu S_x - \mu S_z) + S_z(S_z - \mu S_x - \mu S_y)\right]$$

and $$U = \frac{1}{2E}(S_x^{\ 2} - 2\mu S_x S_y + S_y^{\ 2} - 2\mu S_y S_z - 2\mu S_x S_z + S_z^{\ 2})$$

For axial stress, any two stress values such as S_y and S_z are zero and

$$U = \frac{1}{2E} S_x^{\ 2}$$

For biaxial stresses, any one stress value such as S_z is zero and

$$U = \frac{1}{2E}(S_x^{\ 2} - 2\mu S_x S_y + S_y^{\ 2})$$

An example of the use of strain energy follows. A 1-in. cube of 2024T aluminum alloy is required, in a design situation, to absorb (and release) 4 in.-lb per cu in. of elastic energy. Assuming axial, biaxial, and triaxial equal tensile stresses, what is the maximum value of the stress which can be applied to the cube? $E = 10 \times 10^6$ psi; $\mu = 1/3$.

For axial stress:

$$U = \frac{1}{2E} S^2 \qquad \text{and} \qquad 4 = \frac{1}{2 \times 10 \times 10^6} S^2$$

$$S^2 = 80 \times 10^6 \qquad \text{from which} \qquad S = 8{,}940 \text{ psi}$$

For biaxial stresses:

$$U = \frac{1}{2E}(S_x^{\ 2} - 2\mu S_x S_y + S_y^{\ 2})$$

$$4 = \frac{1}{2 \times 10 \times 10^6}(2S^2 - 2\mu S^2)$$

$$4 = \frac{1}{2 \times 10 \times 10^6}(1.33S^2)$$

and $$S^2 = \frac{80 \times 10^6}{1.33} \qquad \text{from which} \qquad S = 7{,}750 \text{ psi}$$

For triaxial stresses:

$$U = \frac{1}{2E}(3S^2 - 6\mu S^2)$$

$$4 = \frac{1}{2 \times 10 \times 10^6}(3S^2 - 2S^2)$$

and $\quad S^2 = 80 \times 10^6 \quad$ from which $\quad S = 8{,}940$ psi

For a condition of tensile and compressive equal biaxial stresses:

$$U = \frac{1}{2E}(2S^2 + 2\mu S^2)$$

$$4 = \frac{1}{2 \times 10 \times 10^6}(2.67S^2)$$

and $\quad S^2 = \frac{80 \times 10^6}{2.67} \quad$ from which $\quad S = 5{,}480$ psi

1-13. Temperature Effects. When a material is restrained from changing its length due to a change in temperature, stress will be induced in the piece of material. Where the change in temperature is large, and the material is restrained from expanding or contracting, the stress set up is often large enough to cause a permanent deformation in the material. In some cases, failures have occurred from this cause as well as from the fact that ordinary metals are not as strong at high temperatures as they are at normal temperatures. At low temperatures some metals are stronger than they are at normal temperatures.

The coefficient of linear expansion (or contraction) is the linear deformation per inch of length per degree change in temperature. Medium steel has a coefficient of linear expansion of 0.0000067 in. per in. per degree change in temperature Fahrenheit. The coefficients of expansion of different materials are given in the Appendix.

From the definition of the coefficient of expansion, the total deformation caused by a temperature change is determined as follows:

Let

n = coefficient of linear expansion

t = change in temperature, °F

Then

e = coefficient × change in temperature × length

$e = ntL$ = total deformation due to temperature change

From Hooke's law,

$$E = \frac{PL}{Ae}$$

Hence, when the piece of material is restrained,

$$E = \frac{PL}{AntL} = \frac{P}{Ant}$$

from which

$$P = EAnt$$

Thus, the total load induced by a temperature change is dependent on the modulus of elasticity, the area, the coefficient of linear expansion, and the temperature change.

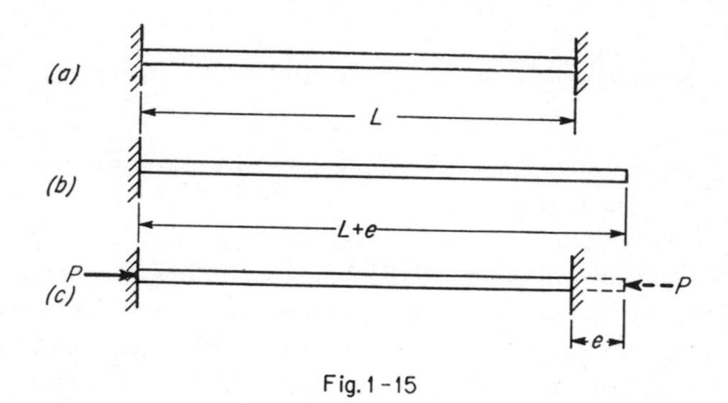

Fig. 1-15

Figure 1-15*a*, *b*, and *c* shows an exaggerated effect of an increase in temperature on a restrained bar. It should be assumed that the bar can support the effect without buckling. Figure 1-15*a* indicates the bar of length L in its restrained position. Figure 1-15*b* shows the same bar with one end released and the increase in length e due to the temperature rise. Because of the restraints at the ends of the bar, the equivalent of a compressive load P (Fig. 1-15*c*) must be applied to the ends to shorten the bar to the original length L, thus inducing a compressive stress in the bar.

Example. A medium-steel tie rod 1 in. $\times$ 2 in. $\times$ 2 ft long has each end fastened in a wall in such a manner that no expansion can take place. If the temperature rises from 20° to 110°F, what compressive load is placed in the rod?

$$
\begin{aligned}
E &= 30{,}000{,}000 \text{ psi} \\
A &= 1 \text{ in.} \times 2 \text{ in.} = 2 \text{ sq in.} \\
n &= 0.0000067 \text{ (from the Table of Coefficients of Expansion in the Appendix)} \\
t &= 110° - 20° = 90° \\
P &= EAnt \\
&= 30{,}000{,}000 \times 2 \times 0.0000067 \times 90 \\
&= 36{,}200 \text{ lb}
\end{aligned}
$$

It is essential to understand the initial conditions of stress in the bar. Then, by the use of the principle of superposition, the additional stress induced by the temperature change can be accounted for. In some

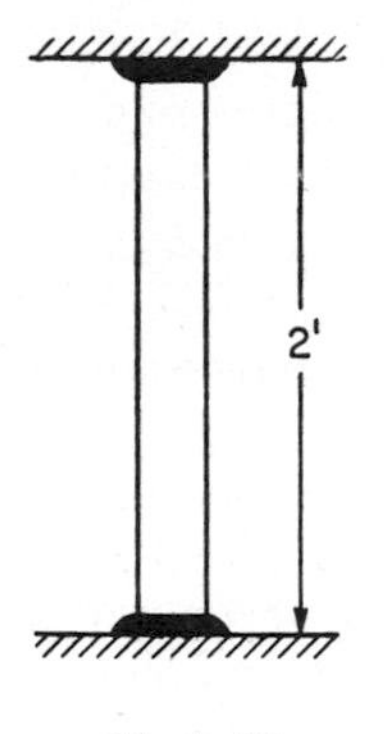

Fig. 1-16

cases, these stresses may neutralize each other and conceivably even reverse the type of stress initially imposed.

A short bar of soft steel is placed as a support and welded as shown in Fig. 1-16. The supports to which the bar is welded are fixed so that the 2-ft. distance between them is constant. The bar is 4 in. × 4 in. in cross section and holds a *tensile* load of 192,000 lb. The temperature of the bar changes from 20° to 100°F. What is the final total load in the bar?

$$\text{Initial load} = 192{,}000 \text{ lb tension}$$

When the temperature increases, the length of the piece must remain 2 ft; hence, the bar effectively contracts by an amount equal to the expansion caused by the increase in temperature.

Since

$$P = EAnt$$

where $n = 0.0000061$ for soft steel

$$P = 30{,}000{,}000 \times 16 \times 0.0000061 \times 80$$

$$= 234{,}200 \text{ lb compression due to temperature change}$$

then

$$\text{Final load} = 234{,}200 \text{ lb compression} - 192{,}000 \text{ lb tension}$$

$$= 42{,}200 \text{ lb compression}$$

Problems

1-1. A steel angle 6 in. × 4 in. × ¾ in. has an allowable stress of 18,000 psi. What axial tensile load can it transmit? See the table in the Appendix for the area of this angle.

1-2. A piece of 2024T aluminum alloy is required to support an axial tensile load of 42,000 lb. Using a factor of safety of 4, what should be the diameter of a round bar to be safe?

1-3. A short yellow-pine post is 4 in. square and 2 ft long. Using a factor of safety of 10, what axial compressive weight W can be applied to the square end of the post?

1-4. In the pin-connected truss shown in Fig. 1-17, determine the required area for each of the members HG, DG, and DF. Use the AISC allowable tensile stress of 22,000 psi.

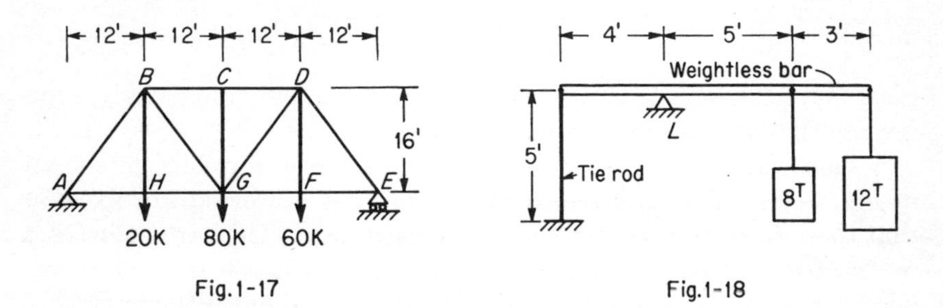

Fig. 1-17

Fig. 1-18

1-5. The 2024T aluminum tie rod and support L (Fig. 1-18) are used to keep the weightless rigid bar in equilibrium. Determine the dimension of the square tie rod using an allowable stress of 18,000 psi.

1-6. In Fig. 1-19, the uniform rigid bar weighs 1,200 lb per ft and is 24 ft long. It is supported by two tie rods as shown, each of which is 10 ft long. AB is made of brass with an allowable stress of 4,800 psi, and CD is made of wrought iron with an allowable stress of 5,400 psi. Determine the required areas for each of the rods.

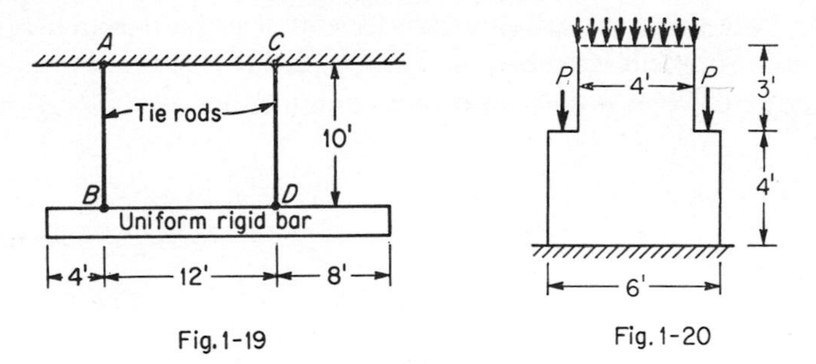

Fig. 1-19 Fig. 1-20

1-7. A solid concrete pier (at 150 lb per cu ft) is uniformly 6 ft thick and has a face area as shown in Fig. 1-20. The top surface is loaded uniformly at the rate of 600 lb per sq ft and each of the equal loads $P = 6{,}000$ lb. Determine the compressive stress in pounds per square inch at the base of the pier.

1-8. The round structural-steel tie rod *FG* in Fig. 1-21 and the smooth hinge at *H* are used to support the weightless 15-ft-long rigid bar. When the load $P = 27$ kips is applied to the bar, what is the required diameter of the rod, using a factor of safety of 3.5?

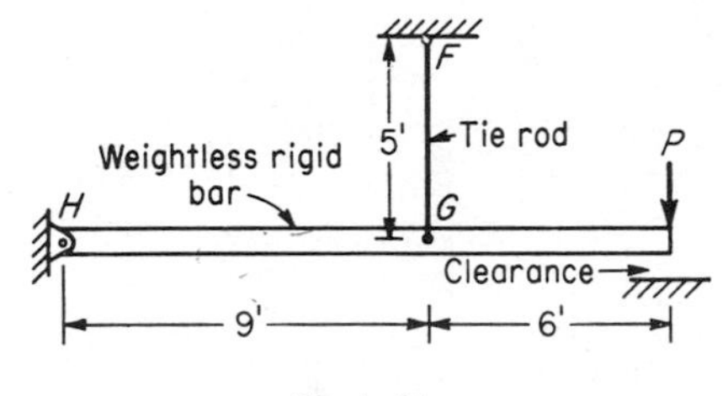

Fig. 1-21

1-9. Determine the required area of the rod *JK* in Fig. 1-22. The material used for the rod is wrought iron, and the design specifies a factor of safety of 3.

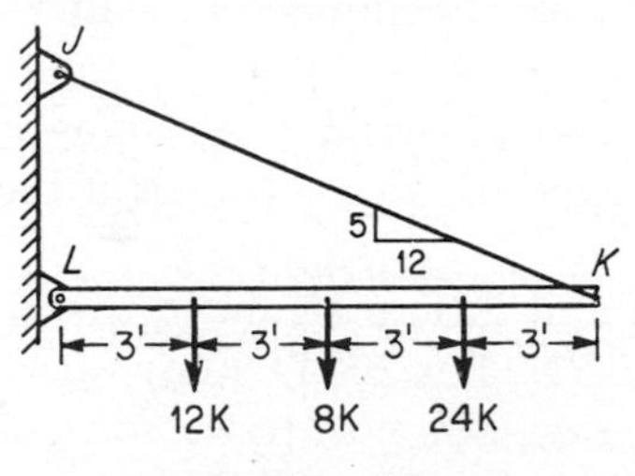

Fig. 1-22

1-10. Determine the areas of bars A and B using design stresses of 16,000 psi for tensile members and 8,000 psi for compressive members (see Fig. 1-23). Assume that any compression member does not buckle.

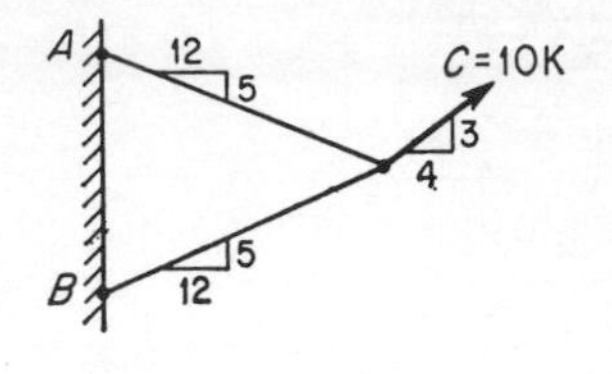

Fig. 1-23

1-11. Figure 1-24 represents a uniform bar of material with cross-sectional dimensions 1.5 in. × 1.2 in. Five axial loads are applied to the bar at sections A, B, C, D, and E. Determine the stress in each of the portions AB, BC, CD, and DE of the bar.

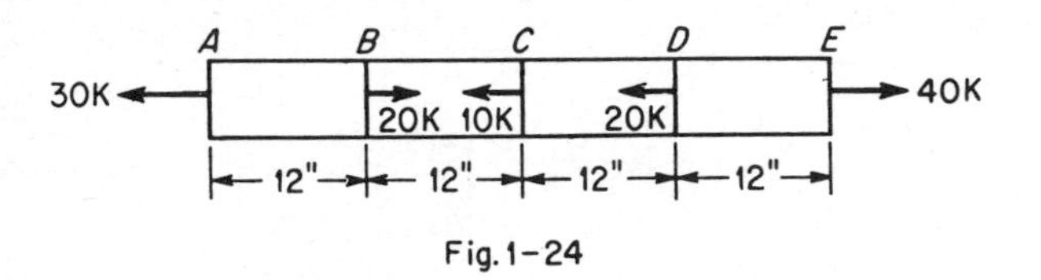

Fig. 1-24

1-12. A 2024T aluminum bar is placed under static axial tensile loading so that the stress developed is 16,000 psi. If the bar is 12 in. long, what is the change in length?

1-13. If the member DF in Prob. 1-4 is made from structural steel, what is its change in length?

1-14. A piece of gray cast iron is 2 in. long and has a diameter of 1.2 in. What applied axial compressive load will decrease its length to 1.99 in.?

1-15. Referring to Prob. 1-5 and its accompanying figure, determine the tangent of the angle of tilt of the rigid bar, assuming that the support L is fixed.

1-16. What is the unit lateral strain in the aluminum-alloy tie rod referred to in Prob. 1-5? For 2024T alloy, $\mu = \frac{1}{3}$.

1-17. A short gray cast-iron post in the form of a hollow tube has an external diameter of 10 in. and a wall thickness of 1 in. It supports a compressive load $P = 600{,}000$ lb distributed over the end of the tube by a heavy plate, which is not shown in Fig. 1-25. What is the unit compressive stress in the cast iron, and what is the factor of safety based on the ultimate strength given in Table 1-1? If the post is 12 in. long, how much is it shortened by the load?

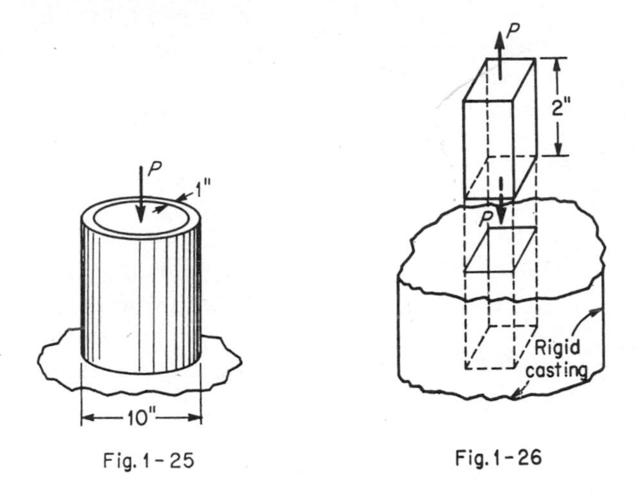

Fig. 1-25 Fig. 1-26

1-18. A 2024T aluminum plug 1 in. × 1 in. × 2 in. long is to be deformed by an axial load P so that it will just fit into a square hole 0.9995 in. × 0.9995 in. in a rigid casting (see Fig. 1-26). Poisson's ratio is 0.333. What tensile force P must be used so that the slip fit can be made?

1-19. What is the total change in length of the bar shown in Fig. 1-24 if the material is structural steel? Refer to Prob. 1-11.

1-20. In Fig. 1-19 and Prob. 1-6, what is the change in length of bar CD, and what is the tangent of the angle of tilt of the rigid bar?

1-21. What must be the minimum clearance as shown in Fig. 1-21 so that the rigid bar does not touch the support when the force P is applied? Refer to Prob. 1-8.

1-22. What is the available unit strain energy in 2024T aluminum alloy if its elastic limit is one-half of its ultimate strength?

1-23. An axial-impact tensile force delivers a total of 22 in.-lb of energy to a 1.5-in. square bar 12 in. long. $E = 10 \times 10^6$ psi. What maximum stress is developed in the bar?

1-24. A 2-in. cube of 2024T aluminum alloy is subjected to biaxial forces $P_x = +28{,}000$ lb and $P_y = -36{,}000$ lb. Determine the total recoverable strain energy. $\mu = 1/3$.

1-25. A bar of material 20 in. long stretches 0.0138 in. under an axial load of 3,450 lb. If the bar has a diameter of 0.505 in. and a diametrical deformation of 0.000101 in., determine (*a*) Poisson's ratio, (*b*) the modulus of elasticity, (*c*) the stress.

1-26. A 2-in. cube of material is subjected to triaxial forces $P_x = -20{,}000$ lb, $P_y = +16{,}000$ lb, and $P_z = -24{,}000$ lb. $\mu = 0.30$. Determine the unit strain in the X-axis direction and the unit strain energy. Use $E = 28 \times 10^6$ psi.

1-27. A steel bar, 36 in. long and 2 in. square, is subjected to an axial energy load of 960 in.-lb. $E = 30 \times 10^6$. What maximum axial stress is developed in the bar? To what axial static tensile force is this stress equivalent? What is the corresponding total stretch?

1-28. What is the total amount of recoverable strain-energy in member *BG* of the truss shown in Fig. 1-17 if the applied loads are removed from the truss? The members of the truss are steel with a design stress of 20,000 psi.

1-29. A bar has a total length of 48 in. and is subjected to an axial tensile load of 30,000 lb. One-half of its length is 1 in. $\times$ 1 in. and the other half is 2 in. $\times$ 2 in. in cross section. $E = 30 \times 10^6$. Neglecting any effect due to the change in the cross-sectional areas, determine the recoverable unit strain energy in each part.

1-30. A ½-in.-diameter and 2-ft-long steel rod is fastened, end to end, to a ½-in.-diameter and 1-ft-long bronze rod. Using $E = 15{,}000{,}000$ psi for the bronze, what is the total change in length of the 3-ft piece when an axial tensile force P of sufficient magnitude to cause a stress of 10,000 psi in the bronze is applied to the piece? What is the recoverable unit strain energy in each part?

1-31. A wrought-iron tie rod 15 ft long and 1 sq in. in cross section is placed in position with no clearance, between the rigid walls, and the nuts are tightened, at 200°F (see Fig. 1-27). When the temperature decreases to 20°F, what is the stress in the rod?

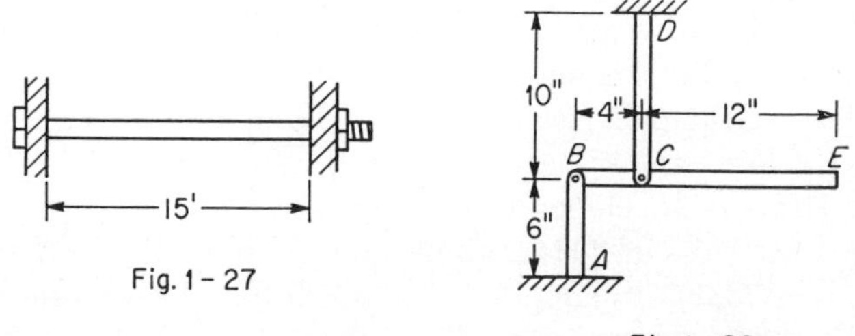

Fig. 1-27

Fig. 1-28

1-32. A steel bar, 0.505 inches in diameter, is suspended vertically from its upper end. If the ultimate strength of the steel is 82,000 psi, what length of bar will just cause it to break? Steel weighs 0.284 lb per cu in.

1-33. If the stretch in a uniform bar of material due to its own weight is given by the equation

$$e = \frac{SL}{2E}$$

what is the total elastic stretch in a bar of steel 0.505 inches in diameter when it is suspended vertically by one end? The elastic limit stress = 41,000 psi.

1-34. For practical purposes, the three bars in Fig. 1-28 can be considered weightless. The pins at B and C are smooth. Bar AB is wrought aluminum and CD is copper. When the temperature of the assembly increases 100°F, what is the total vertical displacement of point E? Neglect the change in length of bar BCE.

1-35. A soft steel bar, 15 ft long with an area of 1.2 sq in., is firmly attached to a rigid wall at each end. The bar sustains a tensile stress of 24,000 psi when it is positioned between the walls. What temperature change will release one-half of the stress?

1-36. The brass and aluminum bars are used to support the rigid weightless bar shown in Fig. 1-29. The temperature of each bar can be controlled independently of each other. Precise measurements of the initial lengths of the bars gives AB = 60.008 in. and CD = 59.996 in. If the temperature of CD is increased by 60°F, what change in temperature in AB will cause the rigid bar to be level?

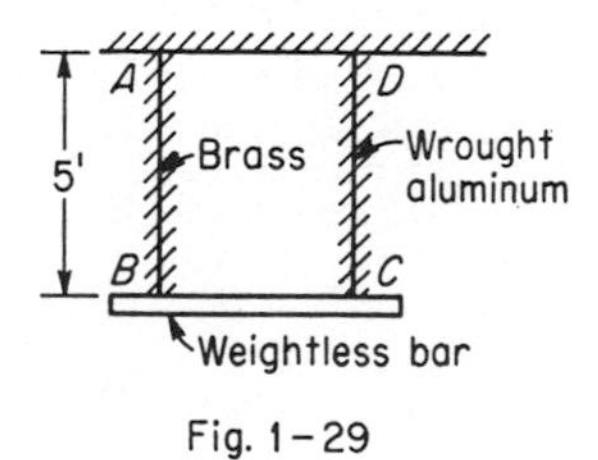

Fig. 1-29

1-37. If the ends of a length of medium-steel rail are the equivalent of being fixed, what stress is developed by a temperature rise of 80°F?

1-38. The change in the volume of a block of material due to the application of an axial tensile force P is given by the equation

$$\Delta V = (1 - 2\mu)\varepsilon V_o$$

where ε is the strain in the direction of the applied force, and V_o is the original volume. (This is developed by continuing the theory used in Arts. 1-9 and 1-10.)

When an axial tensile force of 36,000 lb is applied to opposite surfaces of a 2-in. × 2-in. × 2-in. steel cube for which Poisson's ratio is 0.30, what is the change in the volume of the cube?

1-39. Referring to Prob. 1-38, find the value of Poisson's ratio for a material for which the volume change is zero.

1-40. A cast-iron piece is uniformly 2 in. thick and has the shape shown in Fig. 1-30. It is loaded in compression as shown, where $P = 30{,}000$ lb. Using $E = 12{,}000{,}000$ psi, what is the total shortening in the piece?

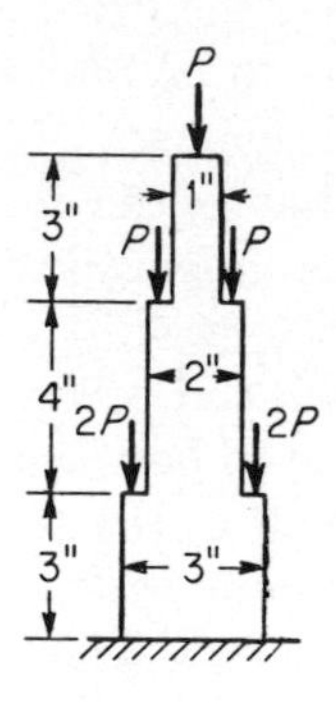

Fig. 1-30

CHAPTER 2
SHEAR AND CONNECTIONS

2-1. Shear Stress. The connection of structural members by the use of rivets, bolts, and welds continues to be of major importance in engineering construction. Because of this importance, their analyses are considered in detail.

When forces are applied parallel to the cross section of a piece of material, one section of the piece tends to slide with respect to the adjacent section, thus causing the material to shear. One of the simplest examples of shearing action is punching a hole in a steel plate.

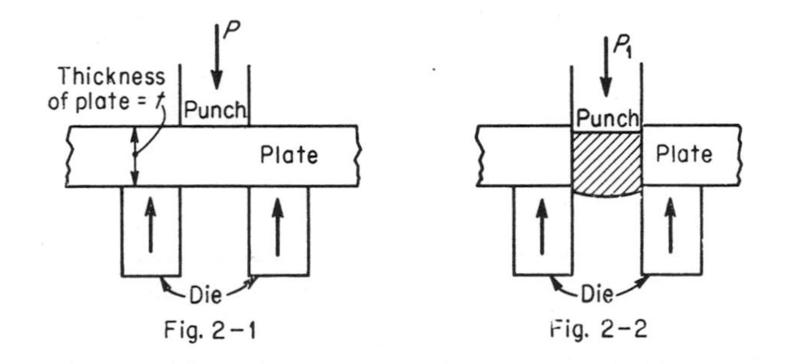

Fig. 2-1 Fig. 2-2

Figures 2-1 to 2-4 progressively show the action and movement which takes place. A force P is applied to the punch, and the die pushes upward on the plate with an equal amount of force. In Fig. 2-2, the shearing action has started, and the force P_1 required will be lessened as there is less metal remaining to be sheared. Figure 2-3, with a still smaller force P_2 required, shows the punch about halfway through the plate. In Fig. 2-4, the shearing action is complete, so P_3 will be a very small amount; enough, however, to overcome the friction between the punch and the plate.

Figure 2-5 shows the piece of the plate (slug) which has been removed. The shaded area is the *area* which has *sheared* and is called the *sheared area.* The student is cautioned against the use of the top circular area of the slug, since no failure, except a compression deformation, has taken place on this face. Hence, the *area* which has failed is the circumferential shaded area equal to the circumference of the slug times the thickness of the plate, or $A_s = \pi \times d \times t$, in which A_s represents the sheared area. (The sub s denotes that the failure was a shear failure.)

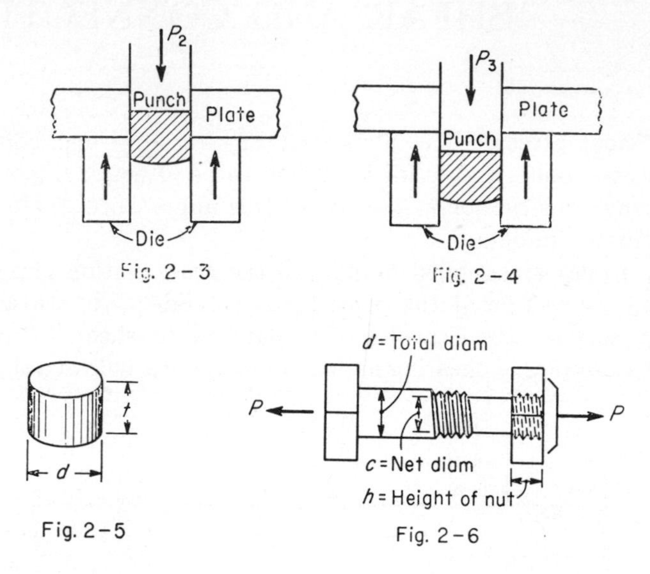

Fig. 2-3 Fig. 2-4

Fig. 2-5 Fig. 2-6

Example. If the ultimate shearing strength of a steel plate is 42,000 psi, what force is necessary to punch a $\frac{3}{4}$-in.-diameter hole in a $\frac{5}{8}$-in. plate?

The area which fails is the circumferential area of the disk to be removed. Hence

$$A_s = \pi dt = \pi \times \tfrac{3}{4} \times \tfrac{5}{8} = \tfrac{15}{32}\pi \text{ sq in.}$$

Then

$$P_s = S_s A_s = 42{,}000 \times \tfrac{15}{32}\pi = 61{,}800 \text{ lb required to punch the hole}$$

Another example of shearing action occurs when the threads are stripped from a bolt. In Fig. 2-6, a nut has partially stripped the threads from the bolt, and careful observation of the smooth shank will show that the *circumferential root area* has failed, leaving the shank threadless. Hence, the sheared area will equal the circumference of the root (net) circle times the height of the nut, or

$$A_s = \pi \times c \times h$$

Example. What force will be required to strip the threads on a $\frac{3}{4}$-in.-diameter bolt when the ultimate shearing strength of the material is 42,000 psi?

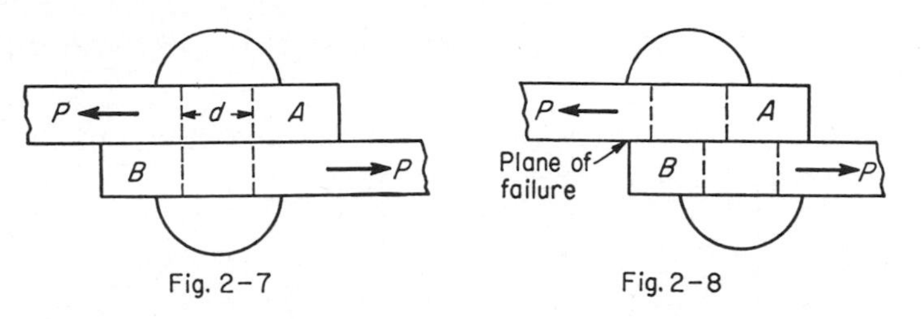

Fig. 2-7 Fig. 2-8

From the Appendix, a $\frac{3}{4}$-in.-diameter bolt has a net (root) diameter of 0.620 in., and the height of the nut (from the table of Bolt Heads and Nuts) is $\frac{3}{4}$ in.

$$P_s = A_s S_s = \pi ch \times S_s = \pi \times 0.620 \times \tfrac{3}{4} \times 42{,}000$$

Then $\qquad P_s = 61{,}300$ lb required to strip the threads

Figures 2-7 and 2-8 show two steel plates held together by a single rivet of diameter d. A force P is transmitted from plate A to plate B by the rivet. In Fig. 2-8, the rivet has been partially sheared because of the applied force P. After the rivet has been completely sheared, plate B is shown in Fig. 2-9, and it should be noticed that the *area* which has *failed* is the *cross-sectional area* of the rivet. Hence

$$A_s = \frac{\pi d^2}{4}$$

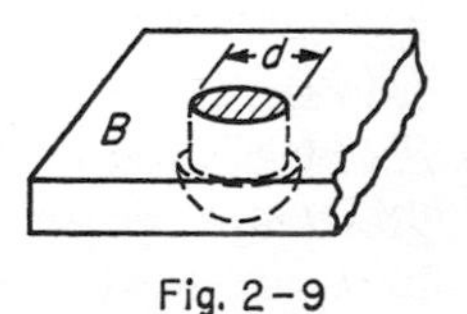

Fig. 2-9

Since only one cross section fails, the failure is called *single* shear.

When two steel plates B and B' are riveted to a single plate A, the force P is transmitted from plate A to plates B and B' (see Figs. 2-10 and 2-11). If the force becomes sufficiently large, the rivet will be sheared simultaneously on *two* different *cross sections*, between plates A and B and between plates A and B'. This failure is called *double shear.* Hence

$$A_s = 2 \times \frac{\pi d^2}{4}$$

These considerations which we have given to rivets hold equally true when bolts are used in place of rivets.

As an example of the shearing action in rivets, determine what force is required to cause a shear failure of a ¾-in.-diameter rivet, as shown in Fig. 2-7. The ultimate shearing stress of the material of which the rivet is made is 42,000 psi.

$$P_s = A_sS_s = \frac{\pi(\frac{3}{4})^2}{4} \times 42{,}000 = 0.4418 \times 42{,}000 = 18{,}560 \text{ lb required}$$

Note. See the Appendix (under Weights and Areas of Bars) to obtain the cross-sectional area of a rivet.

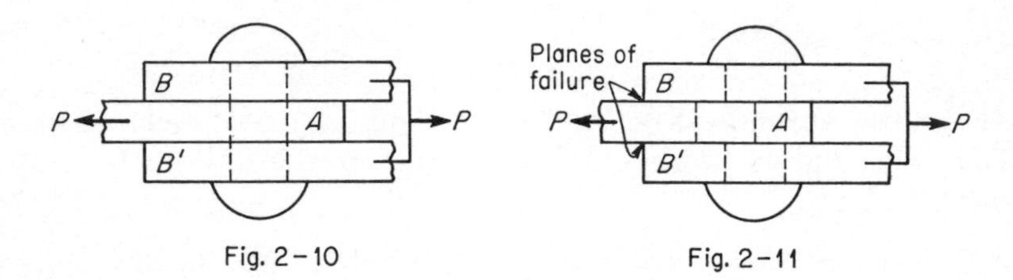

Fig. 2-10

Fig. 2-11

What force will shear a ½-in.-diameter rivet, as shown in Fig. 2-10, when the ultimate shearing strength of the rivet material is 42,000 psi?

$$P_s = A_sS_s = 2\,\frac{\pi d^2}{4} \times S_s = 2\,\frac{\pi(\frac{1}{2})^2}{4} \times 42{,}000$$

$$= 2 \times 0.1963 \times 42{,}000 = 16{,}490 \text{ lb}$$

It should be understood that the illustrative examples solved in the text are based on the supposition of complete failure. In engineering practice, a structure, whether it be a bridge, machine, etc., is designed to be safe against any type of failure, including the limitation of yielding. Hence, the allowable unit stresses used are considerably lower than the ultimate unit stresses which apply when failure occurs.

2-2. Riveted Joints. The primary object of a riveted joint is to transmit force from one steel plate to another, the plates being fastened together by rivets. If the plates are lapped, as shown in Fig. 2-12, the joint is called a *lap joint.*

If the plates are butted together and one cover plate used, the joint is called a *butt joint with one cover plate* (see Fig. 2-13).

When the plates are butted together and two cover plates applied the joint is called a *butt joint with two cover plates* (see Fig. 2-14).

The designation of riveted joints also depends upon the *number of rows* of rivets in a lap joint or the number of rows on the right or left side of a butt joint. Thus, Fig. 2-12 is properly designated a double-riveted lap joint, Fig. 2-13 is a double-riveted butt joint with one cover plate, and Fig. 2-14 is a triple-riveted butt joint with two cover plates.

A row of rivets is a line of rivets in a direction perpendicular to the applied force P.

The *pitch* of the rivets is defined as the distance from the center of one rivet to the center of the next rivet in the same row. The rivets

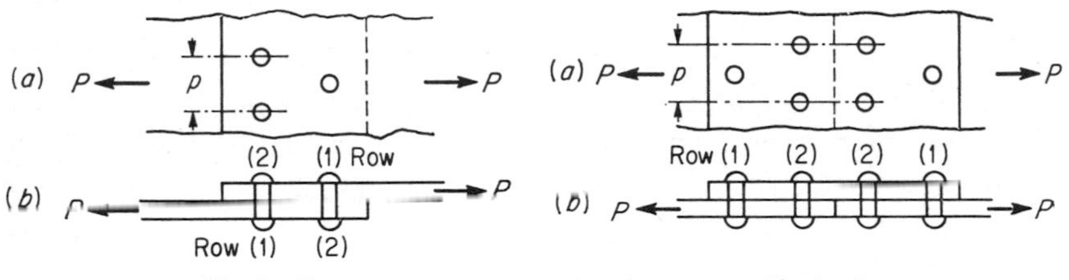

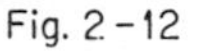

Fig. 2-12

Fig. 2-13

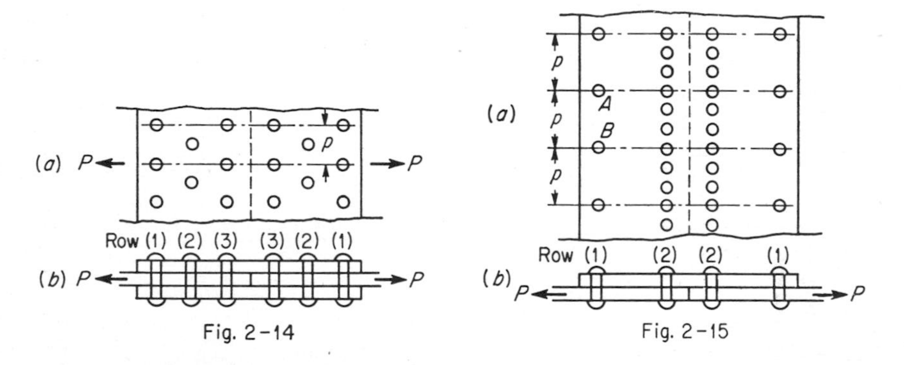

Fig. 2-14

Fig. 2-15

in any row are always spaced equally. When dealing with boiler or pipe joints where the joint is necessarily long, it is customary to use the *longest* distance between the centers of two adjacent rivets in the same row as the pitch, considering such a length of joint as a repeating section. Any number of repeating sections of equal strength may be placed end to end to form the complete joint. Thus, in Fig. 2-15, the repeating section p, of which three are shown, would be used as the basis of the design.

Where the joint is short, or where the total width of the plate is given, it is generally best to consider the entire plate as a unit.

In Fig. 2-12, there are two whole rivets in the pitch distance; in

Fig. 2-13, there are two rivets in the pitch distance; in Fig. 2-14, there are three whole rivets in the pitch distance; in Fig. 2-15, there are four whole rivets in the pitch distance. It is advisable to deal with whole rivets in any pitch distance, although some of the whole rivets are composed of two halves of different rivets in the same row. In Fig. 2-15, the four whole rivets, as stated, consist of one-half of rivet A and one-half of rivet B, both of which are in row 1, and two whole rivets and two half rivets in row 2.

2-3. Riveted Joint Failures. Any riveted joint may fail in one of the following manners:

1. By shearing the rivets (*shear*)
2. By tearing the plate between the rivet holes (*tension*)
3. By crushing the rivets or the plate in front of the rivets, which may be likened to elongating the rivet holes (*bearing*)

2-4. Shear. As the question of rivet shear has been previously discussed, it will suffice to review shear as applied to riveted joints. By referring to Fig. 2-12, we can see that the number of rivets cut off in shear in the pitch distance is two, and each rivet is cut in single shear. Then

$$P_s = 2 \frac{\pi d^2}{4} S_s$$

Also, in Fig. 2-13, two rivets in the pitch distance will be cut off in single shear, and

$$P_s = 2 \frac{\pi d^2}{4} S_s$$

In Fig. 2-14, there are three rivets in the pitch distance but, as each rivet is in double shear,

$$P_s = 6 \frac{\pi d^2}{4} S_s$$

In Fig. 2-15, there are four rivets in single shear in the pitch distance, and

$$P_s = 4 \frac{\pi d^2}{4} S_s$$

If we let n represent the number of rivet areas in the pitch distance, the general equation of shear is

$$P_s = n \frac{\pi d^2}{4} S_s$$

Example. Refer to Fig. 2-16. If the safe shearing stress is 15,000 psi, what safe load can the riveted joint support in shear?

As the plates are 8 in. wide, the total width (length of joint) will be considered as a unit. There are four rivets in single shear on either side of the break in the main plate. Hence

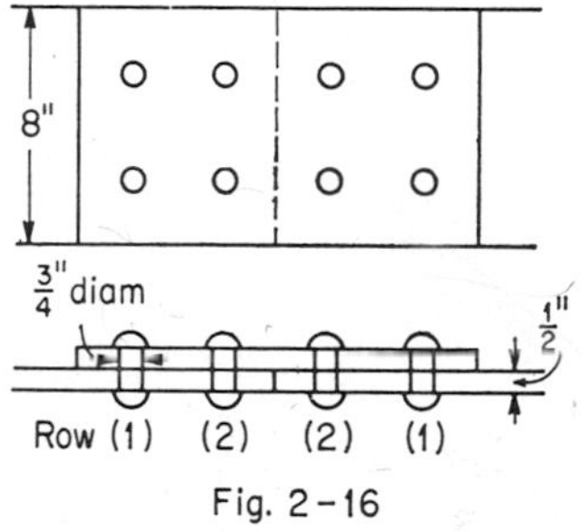

Fig. 2-16

$$P_s = n \frac{\pi d^2}{4} S_s$$

$$= 4 \frac{\pi (3/4)^2}{4} 15{,}000$$

$$= 26{,}500 \text{ lb safe load in shear}$$

2-5. Tension. Tearing the plate between the rivet holes constitutes a tension failure in a riveted joint. Using a single-riveted lap joint as an example, let d = diameter of rivets, t = thickness of plates, and p = pitch distance. Figure 2-17b shows the upper plate of the joint

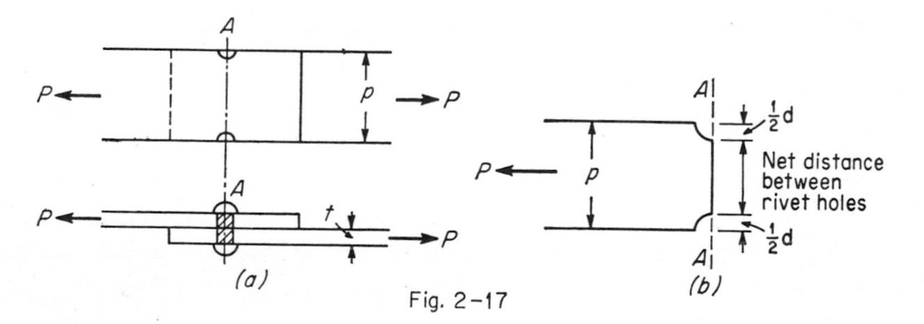

Fig. 2-17

of Fig. 2-17a cut on line AA. The net *distance* between rivet holes is $(p - d)$, and the net *area* between the same holes is $(p - d)t$. It follows that

$$P = AS$$

and

$$P_t = (p - d)tS_t$$

Example. Let it be required to determine the safe tensile load in row 1 of the joint shown in Fig. 2-16 when the safe tensile stress is

18,000 psi. As the total width, 8 in., is considered the pitch distance, then

$$P = 8 \text{ in.} \qquad d = \tfrac{3}{4} \text{ in.} \qquad t = \tfrac{1}{2} \text{ in.} \qquad S_t = 18{,}000 \text{ psi}$$

$$P_t = (p - 2d)tS_t \qquad \text{as two rivet holes occur in row 1}$$

$$= [8 - 2(\tfrac{3}{4})]\tfrac{1}{2}18{,}000$$

$$= (6\tfrac{1}{2})\tfrac{1}{2}(18{,}000)$$

$$= 58{,}500 \text{ lb safe load in tension}$$

2-6. Bearing. The failure due to bearing is one where the rivet or the plate back of the rivet is crushed. If we consider a square pin

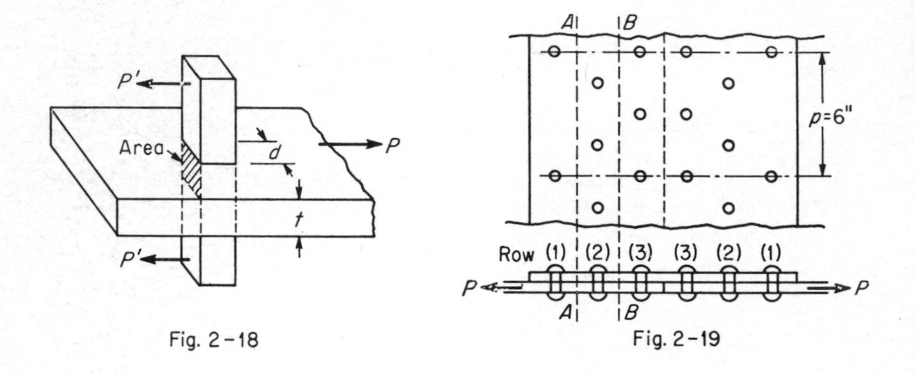

Fig. 2-18 Fig. 2-19

(rivet) inserted in a plate of thickness t (see Fig. 2-18), the area against which the pin pushes is equal to the thickness of the plate times the width d of the pin, or dt. A similar condition prevails if the pin has a circular area, where it is customary to take a section through the center of the rivet perpendicular to the direction of the applied force. Then, this section has a rectangular area of height t and width d, or the area is dt as above.

It should be noted that if the joint or the pitch distance has more than one rivet, all the rivets will be crushed, or all the rivet holes will be elongated.

Example. What is the safe load of the joint shown in Fig. 2-16 in bearing when the safe unit bearing stress is 20,000 psi?

The general formula for bearing stress will be

$$P_b = ndtS_b$$

and as there are four rivet holes to be elongated

$$P_b = 4(\tfrac{3}{4})(\tfrac{1}{2})(20{,}000)$$

$$= 30{,}000 \text{ lb safe bearing load}$$

Example. Figure 2-19 shows a typical boiler joint where the length of the repeating section is 6 in. The rivets are ¾ in. in diameter and the critical main plate is ¾ in. thick. The safe stresses are

$$S_s = 11{,}600 \text{ psi} \qquad S_t = 13{,}750 \text{ psi} \qquad S_b = 22{,}000 \text{ psi}$$

What maximum safe load can be applied to the joint (repeating section)?

Shear:

$$P_s = n\,\frac{\pi d^2}{4}\,S_s$$

There are five rivets in single shear.

$$P_s = 5\,\frac{\pi(\frac{3}{4})^2}{4}\,11{,}600$$

$$= 25{,}600 \text{ lb safe shear load}$$

Tension: considering tensile failure in row 1,

$$P_t = (p - d)tS_t$$

$$= (6 - \tfrac{3}{4})(\tfrac{3}{4})(13{,}750)$$

$$= 54{,}100 \text{ lb safe tensile load on row 1}$$

Bearing:

$$P_b = n\,dtS_b$$

There are five rivet holes to be elongated.

$$P_b = 5(\tfrac{3}{4})(\tfrac{3}{4})(22{,}000)$$

$$= 61{,}900 \text{ lb safe bearing load}$$

It is now necessary to investigate the possible tensile failure in rows other than row 1, when fewer rivets are used in row 1 than in row 2 or 3.

From the results of a number of tests on full-size riveted joints, it has been found safe to assume that each shear area transfers its proportionate part of the total load from the one main plate to the other main plate, in the case of lap joints, or to the cover plates, in the case of butt joints.

Referring to Fig. 2-19, we note that all the load P (per pitch) is applied to, or reaches, the main plate between the rivets in row 1. Since there are five shearing areas to transmit the load P to the cover plate, then

$$P_t = \tfrac{5}{5}(p - d)tS_t = 54{,}100 \text{ lb, as above on row 1}$$

In Fig. 2-20, since *one* rivet area lies between the line AA and the left side of the joint, *one-fifth* of the total load P has been transferred by this rivet area to the cover plate. Hence $\tfrac{4}{5}P$ is left in the main

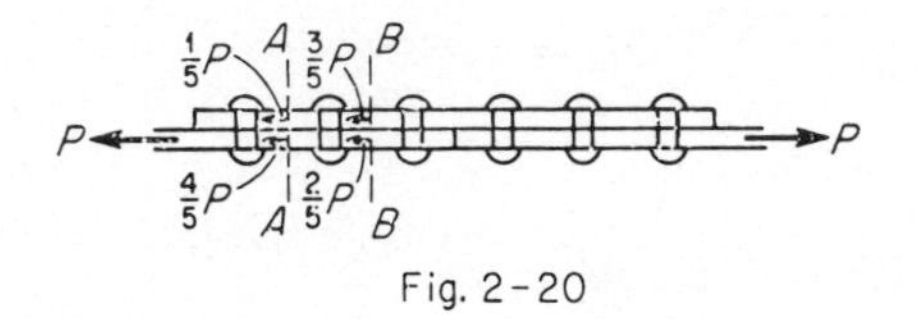

Fig. 2-20

plate and reaches the main plate on the line of rivets in row 2. In this row there is less plate area, because of the two holes passing through the main plate. Hence

$$\begin{aligned} \tfrac{4}{5}P &= (p - 2d)tS_t \\ &= (6 - 2 \times \tfrac{3}{4})\tfrac{3}{4} \times 13{,}750 \\ P_t &= \tfrac{5}{4}(4\tfrac{1}{2})\tfrac{3}{4} \times 13{,}750 \\ &= 58{,}000 \text{ lb safe tensile load on row 2} \end{aligned}$$

In Fig. 2-20, since *three* rivet areas lie between the line BB and the left side of the joint, *three-fifths* of the total load P has been transferred by these rivet areas to the cover plate. Hence $\tfrac{2}{5}P$ is left in the main plate and reaches the main plate on the line of rivets in row 3. In this row there is also less plate area than in row 1, because of the two holes passing through the main plate in row 3. Hence

$$\begin{aligned} \tfrac{2}{5}P &= (p - 2d)tS_t \\ &= (6 - 2 \times \tfrac{3}{4})\tfrac{3}{4} \times 13{,}750 \\ P_t &= \tfrac{5}{2}(4\tfrac{1}{2})\tfrac{3}{4} \times 13{,}750 \\ &= 116{,}000 \text{ lb safe tensile load on row 3} \end{aligned}$$

Analyzing the tensile results for each of rows 1, 2, and 3, it will be seen that

For row 1:

$$P_t = 54{,}100 \text{ lb}$$

when the unit tensile stress is 13,750 psi.

For row 2:

$$P_t = 58{,}000 \text{ lb}$$

when the unit tensile stress is 13,750 psi, or a load $P = 58{,}000$ lb could be used on the joint without producing a unit tensile stress greater than the allowable stress in the main plate on the second line of rivets.

For row 3:

$$P_t = 116{,}000 \text{ lb}$$

which could be used on the joint without producing a unit tensile stress greater than the allowable stress in the main plate on the third line of rivets.

Summarizing, we see that the joint can carry 25,600 lb without exceeding the safe shearing stress in the rivets. While greater loads could be applied to the joint and not exceed the safe bearing or tensile stresses, such as $P_b = 61{,}900$ lb, $P_t = 54{,}100$ lb (row 1), $P_t = 58{,}000$ lb (row 2), $P_t = 116{,}000$ lb (row 3), each one would cause a shearing stress greater than 11,600 psi and would, therefore, not be a safe load for all the stresses. The safe load for the joint is the *least load that it can carry safely*, or in this case 25,600 lb.

It should be noticed that when there is the same number of rivets in each row, the safe load can be calculated for the row in which the main plate transmits the entire load P. Thus, for Fig. 2-16, rows 1 and 2 each contains two rivets; and as the plate in row 1 transmits the entire load P, then P_t for row 1 will be as given (58,500 lb), and P_t for row 2 need not be calculated.

As a further illustration, let us assume that, in Fig. 2-19, there is one rivet in each of rows 1 and 2 and that there are two rivets in row 3, making *four* rivet areas. Then

For row 1:

$$\tfrac{4}{4}P = (p - d)tS_t$$

$$P = \tfrac{4}{4}(6 - \tfrac{3}{4})\tfrac{3}{4} \times 13{,}750$$

$$P_t = 54{,}100 \text{ lb}$$

For row 2:

$$\tfrac{3}{4}P = (p - d)tS_t$$

which need not be calculated, as P is numerically greater than P_t for row 1.

For row 3:

$$\tfrac{2}{4}P = (p - 2d)tS_t$$

$$P_t = \tfrac{4}{2}(6 - 2 \times \tfrac{3}{4})\tfrac{3}{4} \times 13{,}750$$

$$= 92{,}800 \text{ lb}$$

P_b and P_s would be affected, since there are only four rivets; but they have not been calculated, as this illustration is used to emphasize the method of obtaining the safe tensile loads for each row.

For pressure vessels, boilers, and similar applications of riveting, the following recommended stresses will be used in the analyses:

$$S_s = 11{,}600 \text{ psi} \qquad S_t = 13{,}750 \text{ psi} \qquad S_b = 22{,}000 \text{ psi}$$

There are certain refinements which are recommended in the design specifications for pressure vessels, such as the use of the diameter of the rivet holes in determining the bearing areas. These have been purposely omitted from the discussions in this text in order to keep the principles involved straightforward and uncluttered by details.

2-7. Efficiency of Riveted Joints. The efficiency of a riveted joint is defined as the ratio of the strength of the joint to the strength of the unpunched steel plate. If we apply a pull to a piece of steel plate with the same unit width (generally the pitch distance) throughout its length, the plate can fail only in tension, since there are no rivets to be sheared or crushed. The strength (total safe load) of a piece of steel plate then becomes

$$P = ptS_t$$

Then, the efficiency is

$$\text{Efficiency} = \frac{\text{safe load}}{\text{strength of plate}} \times 100\%$$

In the last example, with Fig. 2-19, the safe load was determined as 25,600 lb. The strength of the steel plate is

$$P = 6(\tfrac{3}{4})13{,}750$$

$$= 61{,}900 \text{ lb}$$

$$\text{Efficiency} = \frac{25{,}600}{61{,}900} \times 100 = 41.3\%$$

The ideal riveted joint is one in which the three total loads are equal or nearly so, since the efficiencies of the joint when using P_s, P_t, or P_b will be the same.

2-8. Pressure Vessels. Steam pressure, which tends to rupture a boiler, acts normal to the surface of the boiler just as water pressure does in water pipes. As a result, these pressures must not induce excessive stresses in the material from which the structures are made, or the elastic limit of the materials will be exceeded and failure will ultimately take place.

Figure 2-21 shows a boiler or pressure vessel of diameter D and length L with the longitudinal and circumferential riveted joints. The

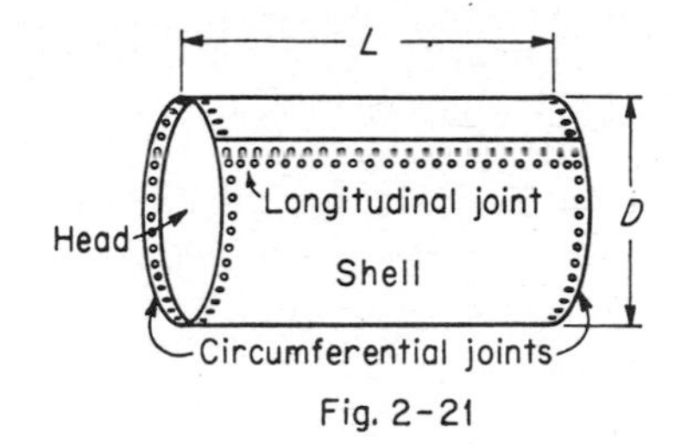

Fig. 2-21

longitudinal joint is used to fasten the shell and is parallel to the geometrical axis of the boiler. The *circumferential* joint is used to fasten the head to the shell.

If we let R represent the internal pressure in pounds per square inch, then the total pressure acting on the head of the boiler is

$$P = AS$$

$$P = \frac{\pi D^2}{4} R$$

Rather than attempt to count the number of rivets in the circumferential joint, it is better to deal with the repeating section of a length equal to the pitch. To calculate the number of pitch distances around the circumference, it is necessary to divide the circumference by the pitch

$$\text{Number of pitch distances} = \frac{\pi D}{p}$$

The pressure which comes to each pitch distance will be the total pressure on the head of the boiler divided by the number of pitch distances, or

$$\text{Pressure per pitch} = \frac{P}{\text{number of pitch distances}}$$

$$= \frac{(\pi D^2/4)R}{\pi D/4} = \frac{RDp}{4}$$

Thus, the load (pressure) per pitch will be

$$P \text{ per pitch} = \frac{RDp}{4}$$

from which the stresses S_s, S_t, or S_b set up in the circumferential joint can be calculated.

Example. A boiler 5 ft in diameter works under a steam pressure of 200 psi. If the pitch of the rivets in the circumferential joint is 3 in., what is the amount of the load per pitch?

Solving the example step by step,
Total load P on the end of the boiler $= AS$

$$P = \frac{\pi D^2}{4} R = \frac{\pi (60)^2}{4} 200$$

$$= 565{,}480 \text{ lb}$$

Number of pitch distances in the circumference

$$= \frac{\pi D}{p} = \frac{\pi 60}{3} = 62.8$$

$$\text{Load per pitch} = \frac{P}{\text{number of pitch distances}}$$

$$= \frac{565{,}480}{62.8} = 9{,}000 \text{ lb}$$

To check by the formula,

$$P \text{ per pitch} = \frac{RDp}{4} = \frac{200 \times 60 \times 3}{4}$$

$$= 9{,}000 \text{ lb}$$

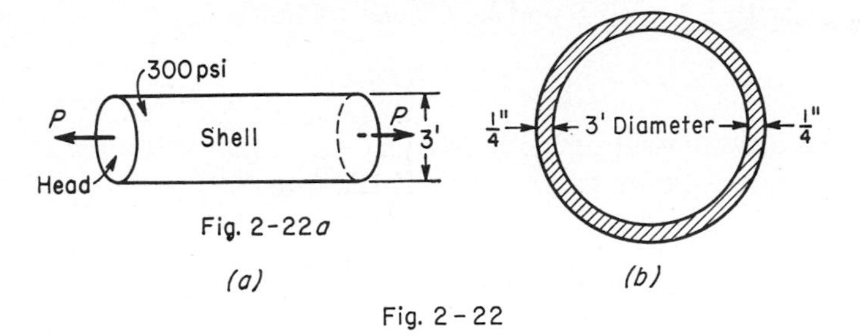

Fig. 2-22

Example. A closed cylinder, 3 ft internal diameter, is made of ¼-in. steel plate (see Fig. 2-22*a*). If the internal pressure is 300 psi, what is the tensile unit stress set up in the circumferential cross section of the shell?

$$P = \frac{\pi D^2}{4} R$$

$$= \frac{\pi (36)^2}{4} 300 = 305{,}360 \text{ lb}$$

Area of the ring of steel (see Fig. 2-22*b*)

$$= \frac{\pi D^2}{4} - \frac{\pi D_1{}^2}{4}$$

$$= \frac{\pi (36.5)^2}{4} - \frac{\pi (36)^2}{4} = 28.47 \text{ sq in.}$$

$$S_t = \frac{P}{A}$$

$$= \frac{305{,}360}{28.47} = 10{,}700 \text{ psi tensile stress in the shell}$$

Figure 2-23 shows a half section of the shell of a thin-walled cylinder, with a length L and a diameter D. The internal pressure R in pounds per square inch acts normal to the shell at every point. As the total pressure P' is the resultant pressure tending to break the shell at AB and CD, it is the sum of the components of all the unit forces in a direction parallel to the resultant. The amount of the resultant pressure P' is obtained by multiplying the unit internal pressure by the projection of the area of the semicircumference $ABECD$ on a plane, which is a rectangle with a length L and height D. Therefore

$$P' = RDL$$

Now, the total pressure P' is resisted by two equal pressures P and P acting on areas AB and CD. Hence, the force

$$P = \frac{P'}{2} = \frac{RDL}{2}$$

As the pitch distance only is the length which is considered, then

$P = RDp/2$ = load per pitch on the longitudinal riveted joint.

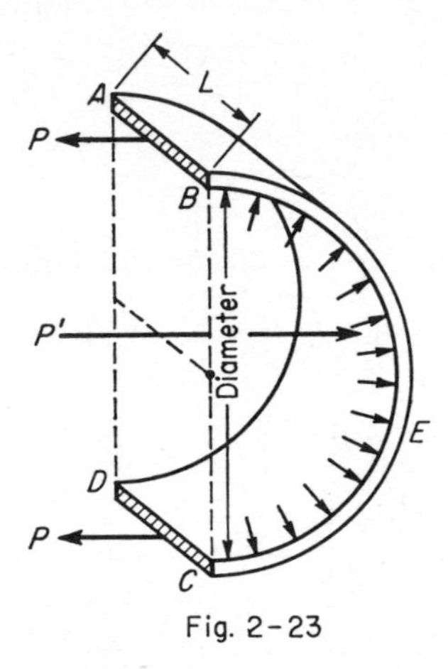

Fig. 2-23

Example. A boiler 5 ft in diameter works under a steam pressure of 200 psi. If the pitch of the rivets in the longitudinal joint is 3 in., what is the total load per pitch?

Solving the example step by step,
Total load P' on projected area of semicircumference $= AR$

$$P' = RDp$$
$$= 200 \times 60 \times 3 = 36{,}000 \text{ lb}$$

Total load P at the AB or CD area (see Fig. 2-23) is $P'/2$

Therefore $$P = \frac{P'}{2} = \frac{36{,}000}{2} = 18{,}000 \text{ lb per pitch}$$

To check by the formula,

$$P = \frac{RDp}{2} = \frac{200 \times 60 \times 3}{2} = 18{,}000 \text{ lb per pitch}$$

With reference to the preceding example and the similar example where the load per pitch on the circumferential joint was solved to be 9,000 lb, it is worth while noting that when the pitch distances on the longitudinal and circumferential joints are equal, *the load per pitch on the circumferential joint is one-half the load per pitch on the longitudinal joint.*

Example. In Fig. 2-22*b*, let it be required to determine the unit tensile stress developed in the longitudinal cross section of the shell. Considering the shell to be 1 ft long (any unit length could be used), then

$$P = \frac{RDL}{2} = \frac{300 \times 36 \times 12}{2}$$

$$= 64{,}800 \text{ lb per ft of length}$$

$$\text{Area } AB \text{ or } CD \text{ (Fig. 2-23)} = 12 \times \tfrac{1}{4} = 3 \text{ sq in.}$$

$$S_t = \frac{P}{A} = \frac{64{,}800}{3} = 21{,}600 \text{ psi}$$

which is the tensile stress on the longitudinal cross section of the shell.

2-9. Structural Riveting. The same general equations apply to structural riveting with certain modifications. The pitch distance with respect to a distance perpendicular to the line of application of the load is not necessary, as all the rivets in the connection are counted.

Example. A 3-in. × 3-in. × ½-in. angle is riveted to a steel plate by four ¾-in. rivets (see Figs. 2-24*a* and 2-24*b*). The plate is ½-in. thick. When $S_s = 15{,}000$ psi, $S_t = 20{,}000$ psi, and $S_b = 32{,}000$ psi, what safe load P may be applied to the angle?
Shear: There are four rivets in single shear. Hence

$$P_s = 4\frac{\pi d^2}{4} S_s = 4(0.442)15{,}000$$

$$= 26{,}520 \text{ lb safe load in shear}$$

Bearing: There are four rivets in bearing. Then

$$P_b = 4dtS_b = 4(\tfrac{3}{4})\tfrac{1}{2}(32{,}000)$$
$$= 48{,}000 \text{ lb safe load in bearing}$$

Tension: from the Appendix, the area of a 3-in. × 3-in. × ½-in. angle section is 2.75 sq in. The net area across any section AA is

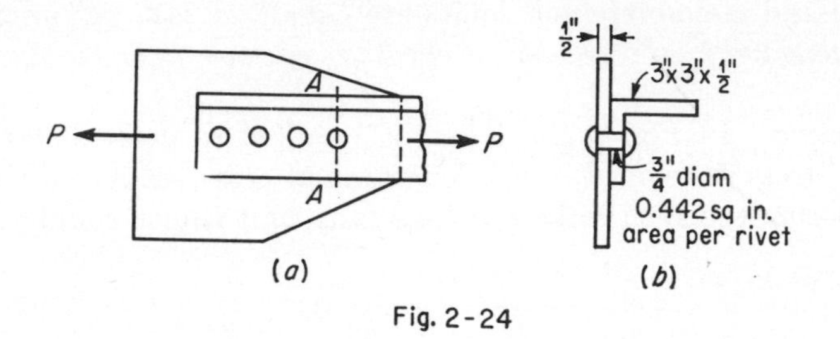

Fig. 2-24

the total area of the angle minus the rectangular cross section through the rivet hole dt, or

$$2.75 - \tfrac{3}{4} \times \tfrac{1}{2} = 2.75 - 0.38 = 2.37 \text{ sq in.}$$
$$P_t = A_tS_t$$
$$= 2.37(20{,}000)$$
$$= 47{,}400 \text{ lb safe load in tension}$$

The maximum safe load which can be applied to the angle is 26,520 lb (shear).

For structural connections, these allowable stresses in pounds per square inch, which are the specifications of the American Institute of Steel Construction, will be used:

$$S_s = 15{,}000 \qquad S_t = 20{,}000$$
$$S_b = 32{,}000 \text{ single shear} \qquad S_b = 40{,}000 \text{ double shear}$$

The design of structural riveted joints includes an additional specification which is not included in the design of pressure-vessel connections. When a rivet is in double shear, as is illustrated in Fig. 2-25 by all the rivets passing through the main plate, the allowable bearing stress is 40,000 psi. For a rivet in single shear, where each rivet passes through

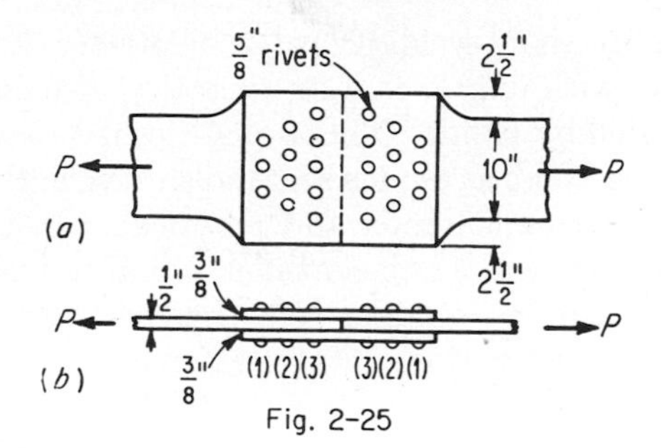

Fig. 2-25

either cover plate, the bearing stress is 32,000 psi. Analyzing the structural tension connection, then

$$P_s = 24 \times 0.307 \times 15{,}000 = 110{,}500 \text{ lb}$$

In the main plate

$$P_b = 12 \times \tfrac{5}{8} \times \tfrac{1}{2} \times 40{,}000 = 150{,}000 \text{ lb}$$

In either cover plate

$$P_b = 12 \times \tfrac{5}{8} \times \tfrac{3}{8} \times 32{,}000 = 90{,}000 \text{ lb}$$

and for two cover plates

$$P_b = 2 \times 90{,}000 = 180{,}000 \text{ lb}$$

For row 1 in tension:

$$P_t = (15 - 3 \times \tfrac{5}{8})\tfrac{1}{2} \times 20{,}000 = 131{,}000 \text{ lb}$$

For row 2 in tension:

$$P_t = \tfrac{24}{18}(15 - 4 \times \tfrac{5}{8})\tfrac{1}{2} \times 20{,}000 = 167{,}000 \text{ lb}$$

For row 3 in tension:

$$P_t = \tfrac{24}{10}(15 - 5 \times \tfrac{5}{8})\tfrac{1}{2} \times 20{,}000 = 285{,}000 \text{ lb}$$

For the narrow plate width:

$$P_t = 10 \times \tfrac{1}{2} \times 20{,}000 = 100{,}000 \text{ lb}$$

From the seven possible answers, the critical capacity of the connection is 100,000 lb.

2-10. Welded Joints. The use of a welding process to join pieces of ferrous metals is not of recent origin. Wrought iron, a historic material, is readily hand-weldable at a relatively low temperature. Low-carbon steel with only a minute quantity of incorporated sulfur is also easily welded by hand. These qualities led to many important engineering devices, such as chains and tools. Now, the use of electric arc welding has become common and practical. In this process, the heat from the arc causes the adjacent base metal to melt and fuse with deposited metal from the welding rod, thus forming the joint. The deposited metal actually transmits the forces from one part being joined to the other part.

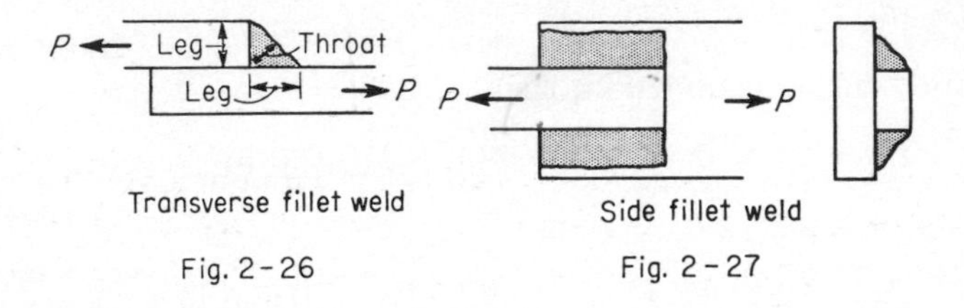

Transverse fillet weld — Fig. 2-26

Side fillet weld — Fig. 2-27

There are two main classes of welds:

1. Fillet welds, shown in Figs. 2-26 and 2-27
2. Butt welds, shown in Figs. 2-28 to 2-30

The plain butt weld is used when the plate thickness is $\frac{3}{8}$ in. or less; the single-V weld is used on plates $\frac{3}{8}$ in. and thicker; the double-V weld is used on thicker plates than the single-V, or where the work can be welded from both sides. The fillet welds and the plain butt welds require no machining before welding, and the fillet welds permit the joining of plates that are to be lapped.

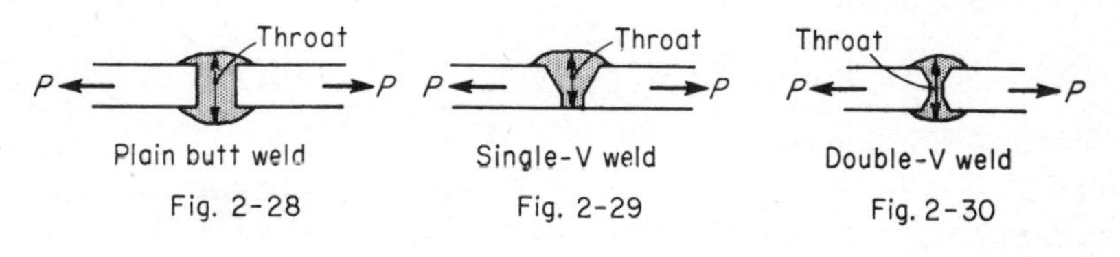

Plain butt weld — Fig. 2-28

Single-V weld — Fig. 2-29

Double-V weld — Fig. 2-30

For fusion welding the American Welding Society gives the following allowable unit stresses for the throat of the welds:

Shear	13,600 psi
Tension	20,000 psi
Compression	20,000 psi

In Figs. 2-28 to 2-30, the throat of the weld has been indicated. However, standard practice requires that the design shall be based on the thickness of the thinnest plate being connected. The following example will show the application of the design method (see Fig. 2-30).

Example. Two ½-in. plates 4 in. wide are to be welded with a double-V weld. What safe tensile load can be used? Since both plates are of equal thickness, and the overrun of the weld material is not considered in making the calculations,

$$\text{Area of the plate} = 4 \times \tfrac{1}{2} = 2 \text{ sq in.}$$

The allowable tensile stress = 20,000 psi.

$$P_t = AS = 2 \times 20{,}000 = 40{,}000 \text{ lb safe load}$$

Since the throat area of a fillet weld is the minimum area through the weld, it becomes the critical or design area. In a fillet weld, shown in Fig. 2-31, the lengths of the two legs are equal, making the cross section of the weld a 45° right triangle. If we assume a ½-in. fillet weld where each leg is ½ in., the throat dimension is

$$\tfrac{1}{2} \text{ in. } \cos 45° = 0.5 \times 0.707$$
$$= 0.3535$$

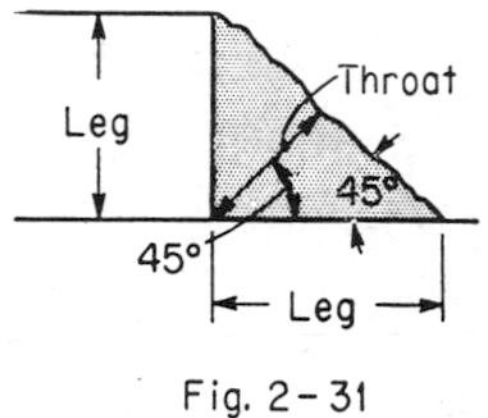

Fig. 2-31

When the area of the throat per lineal inch of weld is

$$0.3535 \times 1 = 0.3535 \text{ sq in.}$$

the shearing value of 1 lineal in. of ½-in. fillet weld is

$$P = AS$$
$$= 0.3535 \times 13{,}600 = 4{,}800 \text{ lb}$$

Similar values for fillet welds of different leg dimensions could be worked but are tabulated as follows:

Leg of fillet, inch	*Allowable shearing load, pounds per lineal inch of fillet weld*
⅛	1,200
¼	2,400
⅜	3,600
½	4,800
⅝	6,000
¾	7,200

Fillet welds placed *either* transverse *or* parallel to the direction of the load will be considered under shear. In Fig. 2-32, welds A and B indicate side fillet welds, and weld C is a transverse fillet weld.

The essential factor in designing welds to resist or transmit forces is to have the center of gravity of the lengths of the welds coincide with the line of action of the forces.

Example. Two ¼-in. plates, as shown in Fig. 2-32, are to be fillet-welded with ¼-in. fillets. If $P = 31{,}200$ lb, design the welded connection. Allowable shearing load per lineal inch of ¼-in. fillet

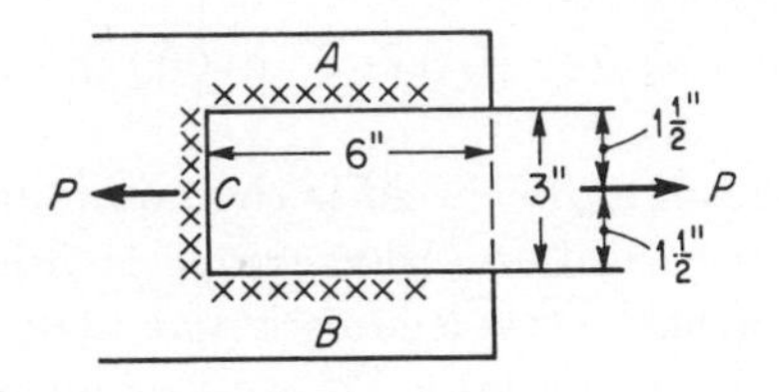

Fig. 2-32

weld = 2,400 lb. Let L = length of weld in inches.

$$L = \frac{P}{2{,}400} = \frac{31{,}200}{2{,}400} = 13 \text{ in.}$$

The design requires that 13 lineal in. of welding be used, which should be located so that its center of gravity coincides with the line of action of the force P. As the two sides A and B cannot provide more than 12 lineal in., it is necessary to weld along line C, which provides for 3 in., and divide the remaining length equally along A and B. Then

$$A + B = 13 - C = 13 - 3 = 10 \text{ in.}$$

$$A = B$$

So $\quad 2A = 10$ in. $\quad$ and $\quad A = B = 5$ in. long

If more welding is necessary, up to 3 in. of additional welding can be added on the underside of the 3-in. plate at the end of the wider plate.

Example. A 4-in. × 3-in. × ½-in. angle section is to have the 4-in. leg fillet-welded, with ½-in. fillet welds, to a steel plate. Using side welds only, find the length of weld necessary and give the lengths A

and B in Fig. 2-33 when $P = 43{,}200$ lb. From the Appendix, the centroid of a 4-in. × 3-in. × ½-in. angle along the 4-in. leg is 1.33 in. from the corner of the angle. Allowable shear per inch of ½-in. fillet weld is 4,800 lb. Length L of weld required:

$$L = \frac{P}{4{,}800} = \frac{43{,}200}{4{,}800} = 9 \text{ in.}$$

Now this 9-in. length must be divided into lengths A and B so that their center of gravity coincides with the line of the load P. Using a

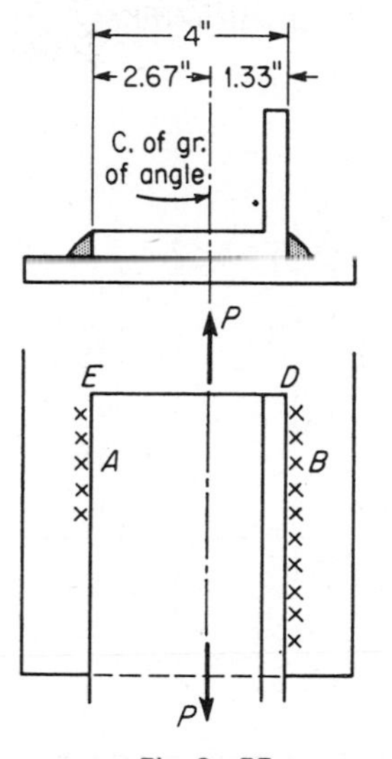

Fig. 2-33

force-summation equation where P equals the total resisting forces in the side fillet welds,

$$P = 4{,}800 \text{ lb/in. } (A \text{ in.} + B \text{ in.})$$

Dividing both sides of the equation by 4,800 lb per in., the equation reduces to

$$9 \text{ in.} = A \text{ in.} + B \text{ in.}$$

Writing a moment equation (using lengths of welding) about the toe "E" of the angle,

$$A \times O + B \times 4 - 9 \times 2.67 = 0$$

$$B = \frac{9}{4} \times 2.67 = 6 \text{ in.}$$

Writing a moment equation about the heel "D" of the angle,

$$B \times O - A \times 4 + 9 \times 1.33 = 0$$

$$A = 3 \text{ in.}$$

Checking the results of the moment equations,

$$9 \text{ in.} = 3 \text{ in.} + 6 \text{ in.}$$

Example. In the preceding problem, let it be assumed that length B cannot exceed 5 in. Then, it is necessary to weld along side C,

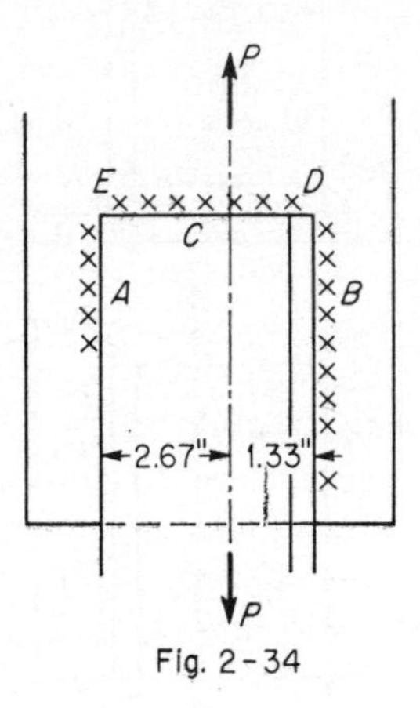

Fig. 2-34

which is 4 in. (see Fig. 2-34); but, in doing this, the center of gravity of the A, B, and C welds must be kept on the line of the load P. Now

$$A + B + C = 9 \text{ in.}$$

$$A + B + 4 = 9 \text{ in.}$$

$$A + B = 5 \text{ in.}$$

Writing a moment equation about the toe "E" of the angle,

$$A \times O + B \times 4 + C \times 2 - 9 \times 2.67 = 0$$

$$B \times 4 + 4 \times 2 - 9 \times 2.67 = 0$$

$$4B = 24 - 8$$

$$B = 4 \text{ in.}$$

Writing a moment equation about the heel "D" of the angle,

$$B \times O - A \times 4 - 4 \times 2 + 9 \times 1.33 = 0$$

$$-4A = -12 + 8$$

$$A = 1 \text{ in.}$$

Checking,

$$A + B + C = 9 \text{ in.}$$

$$1 + 4 + 4 = 9$$

Problems

2-1. An aluminum-alloy plate ½ in. thick is to have a 13⁄16-in.-diameter hole punched in it. If 52,400 lb is required to punch the hole, what is the ultimate unit shearing stress of the material?

2-2. Two plates, as specified in Prob. 2-1, are to be riveted together with a ¾-in.-diameter aluminum-alloy rivet. What safe load could the rivet transmit when the allowable unit shearing strength of the 2024T aluminum alloy is 9,000 psi?

2-3. Two ⅜-in. plates are to be riveted to a ½-in. plate by two ⅝-in.-diameter rivets. All material is 2024T aluminum alloy. Based on shearing action and a factor of safety of 4, what load can be applied?

2-4. Figures 2-35*a* and 2-35*b* show a steel plate X bolted by a ¾-in.-diameter bolt to two steel angles. The angles are riveted to a steel plate Y by two ⅞-in.-diameter rivets. If the rivets are subjected to tensile stress only, what safe load P may be applied, based on a safe stress of 18,000 psi?

2-5. Refer to Figs. 2-35*a* and 2-35*b*. If the bolt fails by shear on the shank area, what load P may be applied? Use a safe stress of 20,000 psi for the high-strength bolt material. Comparing the results

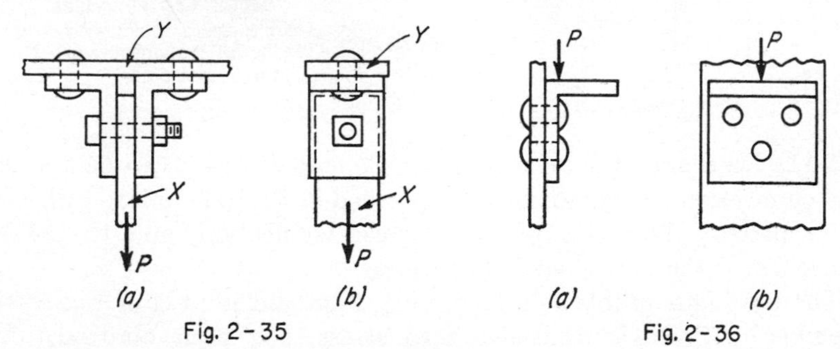

Fig. 2-35

Fig. 2-36

of Probs. 2-4 and 2-5, what safe load would you recommend for the connection?

2-6. A steel angle is riveted by three rivets, each ¾-in. diameter, to a steel plate, shown in Figs. 2-36*a* and 2-36*b*. If the rivets have an allowable shear stress of 20,000 psi, what would be the recommended safe load P?

2-7. Aluminum-alloy rivets are used to connect two plates, shown in Fig. 2-7. The rivets are ⅝-in. diameter and have a safe shear stress of 10,000 psi. What is the safe load when seven rivets are used to make the connection?

2-8. A ¾-in.-diameter steel bolt with a nut is tightened until the tensile force is 32,000 lb. Will the bolt fail, either in tension or by stripping the threads, if the ultimate tensile strength of steel is 100,000 psi and the ultimate shearing strength is 48,000 psi?

2-9. A 4-in. × 6-in. × ¾-in. steel angle is riveted by four ¾-in. rivets to the steel column shown in Fig. 2-37. Determine the maximum force P which can be applied. The tension analysis does not apply.

2-10. Two steel plates are butted together and connected by a single cover plate. All material is 14 in. wide and ½ in. thick. Eight ⅞-in. bolts arranged in two rows of four bolts each are used on each side of the joint. Determine the maximum tensile load which can be applied to the structural connection.

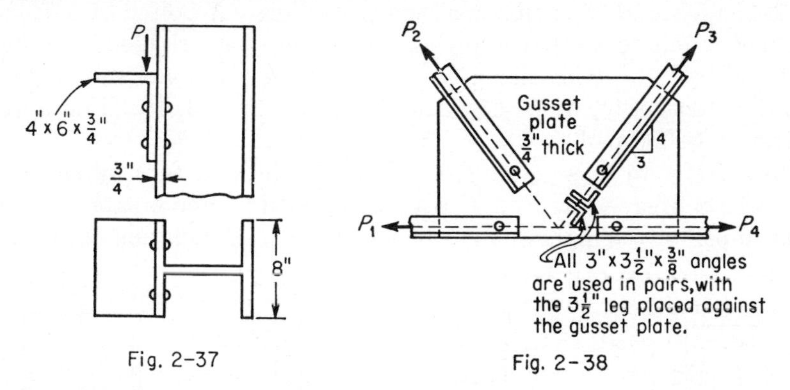

Fig. 2-37

Fig. 2-38

2-11. Determine the maximum tensile load P which can be applied to a connection similar to the one specified in Prob. 2-10 but with two cover plates. This is a pressure vessel connection, and the 14-in. dimension is the unit length of the joint.

The next four problems refer to Fig. 2-38 and joint G of Fig. 1-17. Member CG, which transmits zero stress, has been omitted. An unknown number of ¾-in.-diameter rivets will be required to connect each pair of angles to the gusset plate. (Only one rivet is shown for each member.)

2-12. Using the maximum of the four forces at joint G, determine the maximum tensile stress in the angles.

2-13. Considering shear only, how many rivets are required to connect the P_2 angles to the gusset plate?

2-14. Considering bearing (single shear) only, how many rivets are required to connect the P_4 angles to the gusset plate? How many rivets are required to make this same connection if shear stress is the basis of design?

2-15. Assuming that the same type of stress (shear or bearing) as has been found critical in Prob. 2-14 is also critical in the other members, determine the required number of rivets for the P_1 and P_3 angles.

2-16. In Fig. 2-39, the length of the repeating section is 12 in. What safe load will the joint stand in shear, in tension, and in bearing? Investigate for tension in rows 1 and 2. (Pressure vessel connection.)

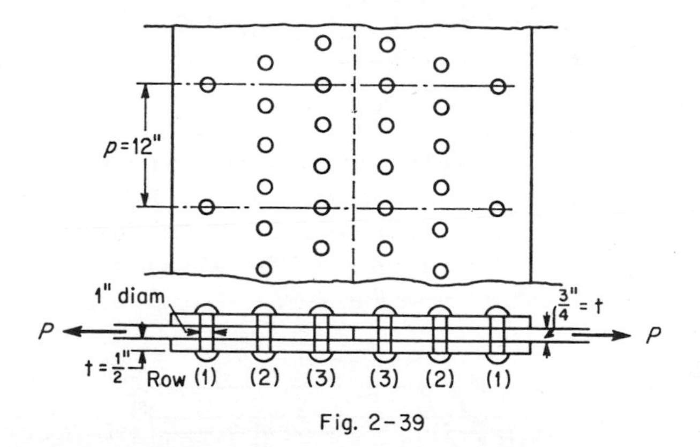

Fig. 2-39

2-17. Referring to Prob. 2-16 and Fig. 2-39, determine the efficiency of the joint.

2-18. A nonsymmetrical structural connection shown in Fig. 2-40 is made of a ½-in. main plate and ⅜-in. cover plates. The rivets are ⅞-in. diameter. Determine the maximum tensile capacity P of the connection.

2-19. How efficient is the connection of Prob. 2-18?

2-20. A pinned structural anchorage consists of the 2-in.-diameter steel pin and the three pieces of steel plate. No allowance will be made for friction between the plates. Determine the optimum dimensions

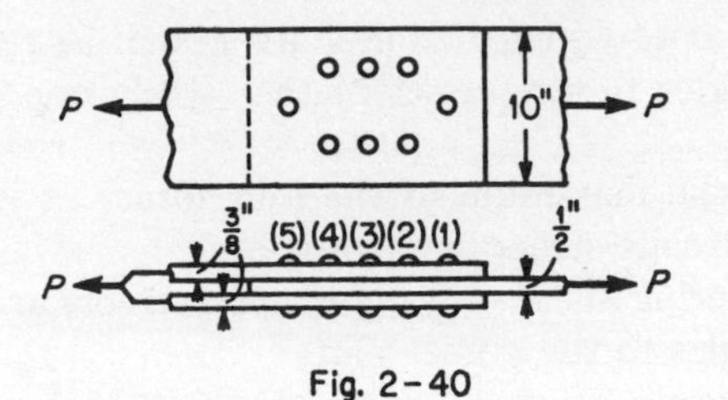

Fig. 2-40

Y and *Z* consistent with the shear and/or bearing strength of the connection shown in Fig. 2-41.

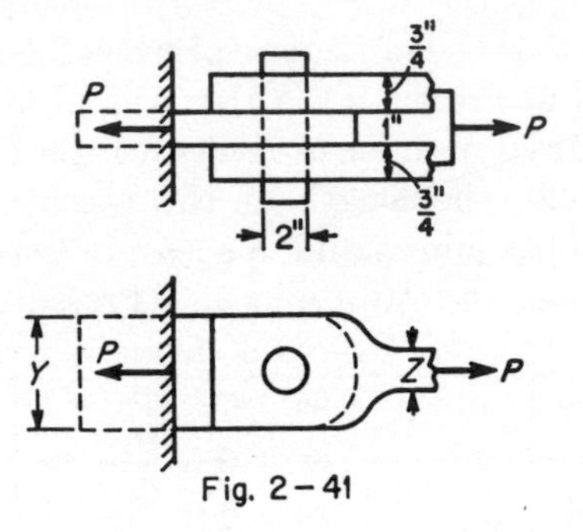

Fig. 2-41

2-21. Two 1-in.-diameter bolts connect the two pieces of 2-in. × 6-in. full-size timber to the 1-in. × 6-in. steel plate. The minimum length *L* from the center of the bolts to the end of the timber is 5 in. If the safe shear strength of the timber is 100 psi, what tensile stress is developed in the timber? See Fig. 2-42.

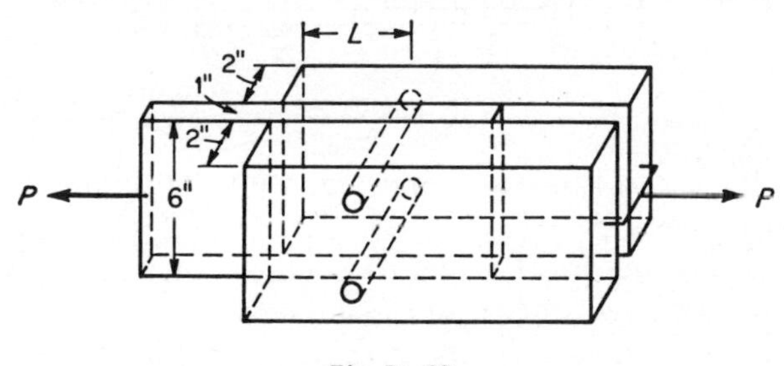

Fig. 2-42

2-22. What is the pressure per 6-in. pitch on the longitudinal joint of a 6-ft-diameter boiler working under a steam pressure of 300 psi?

2-23. In Prob. 2-22, what is the pressure per 3-in. pitch on the circumferential joint of the same boiler?

2-24. The longitudinal boiler joint of a 5-ft-diameter boiler is like the joint in Fig. 2-19. The steam pressure is 150 psi. Calculate the unit shearing stress and the unit bearing stress in the rivets. What is the unit tensile stress in the plate in row 1?

2-25. Figure 2-13 shows the repeating section of the circumferential joint on a 3-ft-diameter boiler with 180 psi steam. Pitch = 3 in.; diameter of rivets = ½ in.; t = ⅜ in. What are the values of S_s, S_t, and S_b in the joint?

2-26. A boiler 4½ ft in diameter is made of ½-in. steel plate. The longitudinal joint is a double-riveted butt joint with two cover plates, one rivet in each row in the pitch distance. Rivets are ¾ in. in diameter at 3-in. pitch. What maximum steam pressure can be used for this pressure vessel?

2-27. The circumferential joint of the boiler in Prob. 2-26 is a double-riveted lap joint with ⅝-in. rivets at 3-in. pitch. Using the allowable stresses, what maximum steam pressure can be used? There is one rivet in each row in the pitch distance.

2-28. A ⅜-in. steel plate 2 in. wide is to be lapped on a 4-in. wide steel plate and welded to it by ⅜-in. side fillet welds. What minimum length of weld on each side of the 2-in. plate can be used if the load applied, as shown in Fig. 2-32, is 54,000 lb?

2-29. If the 2-in. plate in Prob. 2-28 is to have a ⅜-in. transverse fillet weld in addition to the side welds, what minimum length of each side weld can be used?

2-30. A double-V weld is to be used to connect two 3-in. × ¾-in. steel plates. Assuming no overrun of welding material, what total tensile load may be safely applied to the plates?

2-31. A 4-in. × 4-in. × ½-in. angle is to be side-welded by ½-in. fillet welds to a steel plate. The load applied to the center of gravity of the angle section is 72,000 lb. What lengths of side welds will be necessary?

2-32. A 3½-in. × 4-in. × ½-in. angle is to have the 3½-in. leg welded to a plate by ½-in. transverse and side fillet welds. The load to be carried is 80,000 lb applied to the center of gravity of the 3½-in. leg. Calculate the side lengths of the welds.

2-33. What safe load may be applied to the welded joint shown in Fig. 2-43?

2-34. Design ⅜-in. side fillet welds to fasten the angles designated by P_2 to the gusset plate shown in Fig. 2-38.

2-35. Design ⅜-in. side and transverse fillet welds to fasten the angles designated by P_4 to the gusset plate shown in Fig. 2-38.

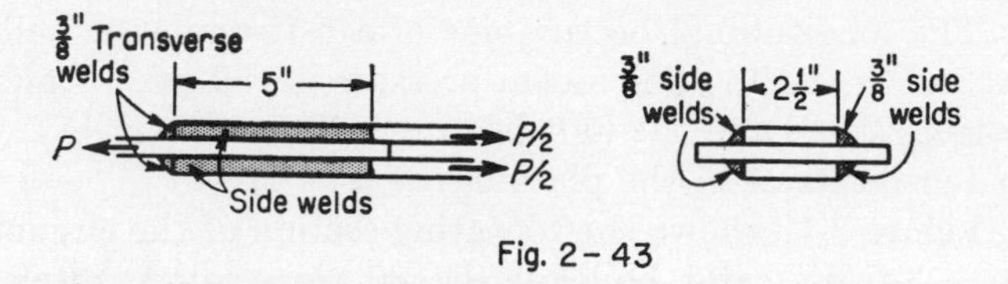

Fig. 2-43

2-36. A pair of angles of the same size shown in Fig. 2-38 are to be stressed to their maximum tensile capacity of 20,000 psi. Design ⅜-in. side fillet welds to join them to a gusset plate.

2-37. Design ⅜-in. side and transverse fillet welds for a pair of the same angles shown in Fig. 2-38. The angles are to be stressed to their maximum tensile capacity of 20,000 psi.

2-38. Figure 2-44 shows the cross section of a 30-in.-diameter pipe made of ½-in. plate. A single-V weld is used on the longitudinal seam. What steam pressure may be safely used in the pipe? Assume a unit length of pipe of 1 ft, and assume that the weld will fail in tension under the action of the steam.

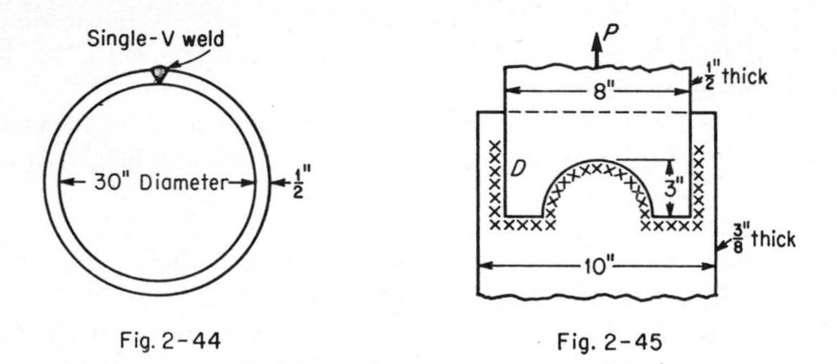

Fig. 2-44 Fig. 2-45

2-39. Assume that the angle shown in Fig. 2-37 is to be welded at the heel and toe to the steel column by ¾-in. (transverse) fillet welds. What force P would you recommend?

2-40. The welded lap joint shown in Fig. 2-45 is to be designed for the maximum tensile capacity of its plates at 20,000 psi. A ⅜-in. fillet weld is to be used. Determine the length of each side fillet "D."

CHAPTER 3
TORSION

3-1. Torque. In the previous chapters, the effects of the applications of axial, biaxial, and triaxial forces were emphasized. These effects were studied from the basic considerations of stress, strain, and strain energy.

There are numerous examples in engineering practice where systems of couples, which cause a turning effect, are applied to pieces of deformable materials. In this chapter, the study is directed solely to couples which result in a rotating or turning effect. Figure 3-1 shows a *couple*, consisting of two equal, opposite, noncollinear, parallel forces applied to a pulley. The couple is attached to a cylindrical shaft.

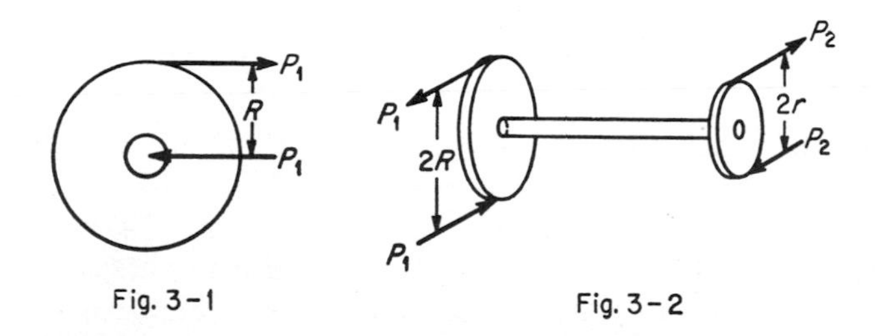

Fig. 3-1 Fig. 3-2

The moment or *torque* of the couple is equal to the product of either force P_1 and the perpendicular distance R between them. Thus

$$T = P_1 R$$

where T is the external torque, which is usually expressed in inch-pound units.

A quick review of the stress-analysis theory in the preceding chapters will remind us that the material part is always in a condition of static equilibrium. Hence, the external torque is balanced by another external couple or by an internal torque created by the internal resisting stresses. Figure 1-7 illustrates the axial condition of equilibrium. A possible torque condition of equilibrium is shown in Fig. 3-2, where

$$T_1 = P_1 2R \qquad \text{and} \qquad T_2 = P_2 2r$$

are equal in magnitude, but have opposite senses of turning. Then

$$T_1 - T_2 = 0 \qquad \text{or} \qquad T_1 = T_2$$

3-2. Shearing Stresses. The result of the application of a torque to a shaft is clearly shown in Fig. 3-3. The line AB on the shaft was straight before the external torque was applied. The left couple, $P2R$, is supplied by the wall and prevents the left end of the shaft from rotating. The right couple, also $P2R$, causes a twist to develop in the

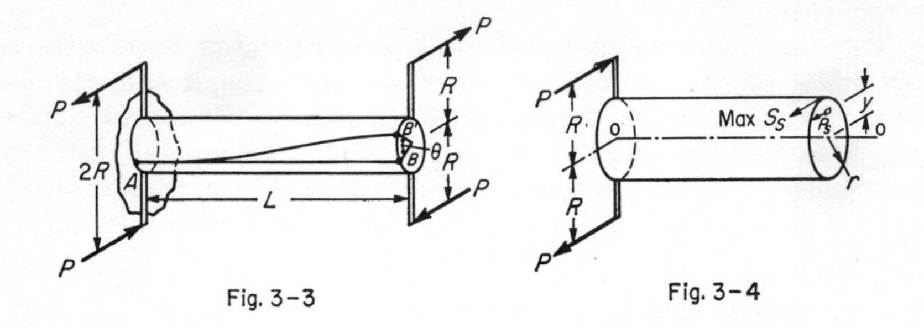

Fig. 3-3

Fig. 3-4

shaft which is shown by the new position AB' of the initially straight line AB.

Hooke's law states that within the elastic range of stress, the ratio of stress to strain is a constant. The law also implies that strain is always accompanied by a corresponding stress. Figure 3-4 shows *one* of a large number of internal resisting shearing forces P_s. Since we know that the external and internal torques must be equal,

$$T = P2R = \Sigma P_s Y$$

where Σ (sigma) denotes the sum of all the internal torques.

Certain basic assumptions are made and have been proved true by tests. These assumptions are:

1. A plane (cross) section remains plane.

2. Shearing strain varies directly as the distance from the central axis O-O.

3. Shearing stress is proportional to shearing strain.

Then the maximum shearing stress S_s occurs at the surface of the shaft such that

$$\frac{S_s}{r} = \frac{S_s'}{y}$$

where S_s' is the shearing stress at a distance y from the central axis.

Assuming that the S_s' stress acts on a small particle of the area dA, then $P_s = S_s' dA$ and from above

$$T = \Sigma S_s' y \, dA$$

From

$$\frac{S_s}{r} = \frac{S_s'}{y}$$

$$S_s' = \frac{S_s y}{r} \qquad \text{and} \qquad T = \frac{S_s}{r} \Sigma y^2 \, dA$$

By the use of advanced mathematics which evaluates the summation process,

$$T = \frac{S_s J}{r} \qquad \text{or} \qquad T = \frac{S_s' J}{y}$$

where T = external torque, in.-lb
S_s = maximum shearing stress
S_s' = shearing stress at a distance y from the central axis
J = polar moment of inertia of the cross section of the shaft (see Table 3-1)
r = external radius, in.
y = distance from the central axis (in.) to the point where S_s' is to be calculated

TABLE 3-1

Type of shaft	Value of J
Solid circular	$\frac{\pi r^4}{2}$
Hollow circular	$\frac{\pi}{2}(r^4 - r_1^4)$

Example. A solid circular steel shaft 2 in. in diameter is twisted by a torque = $20P$ (inch-pounds). The allowable unit shearing strength of the steel is 12,000 psi. What force P can be applied?

$$T = \frac{S_s J}{r}$$

From Table 3-1

$$J = \frac{\pi r^4}{2}$$

Hence $$P \times 20 = \frac{12{,}000 \times \pi(1)^4/2}{1}$$

from which $$P = \frac{12{,}000 \times \pi(1)^4}{20 \times 1 \times 2} = 942 \text{ lb}$$

Example. A hollow shaft is twisted by a couple consisting of 4,500 lb-forces as shown in Fig. 3-5. What is the maximum unit shearing stress in the hollow shaft?

For a hollow shaft

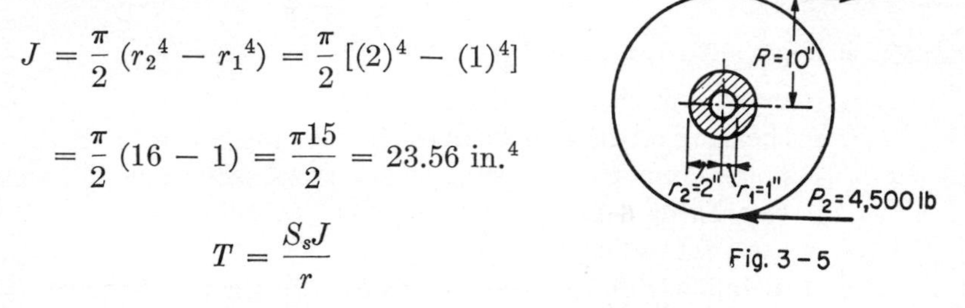

Fig. 3-5

$$J = \frac{\pi}{2}(r_2^4 - r_1^4) = \frac{\pi}{2}[(2)^4 - (1)^4]$$

$$= \frac{\pi}{2}(16 - 1) = \frac{\pi 15}{2} = 23.56 \text{ in.}^4$$

$$T = \frac{S_s J}{r}$$

$$T = 2PR = 2(4{,}500)10 = 90{,}000 \text{ in.-lb}$$

Then $$90{,}000 = \frac{S_s \times 23.56}{2}$$

from which $$S_s = \frac{90{,}000 \times 2}{23.56} = 7{,}640 \text{ psi}$$

The shearing stress, as has been noted, is a maximum at the surface of the shaft and is zero at the center. This stress increases uniformly from the center to the outside surface, so that the stress is proportional to the distance from the center. An example will illustrate this fact. In the preceding numerical example, the stress at the outside surface was calculated to be 7,640 psi at a distance of 2 in. from the center. If the stress at the inside surface (at a distance of 1 in. from the center) is desired, then with S_s' representing the required stress

$$\frac{S_s' \text{ at the inside surface}}{1} = \frac{S_s \text{ at the outside surface}}{2}$$

$$\frac{S_s'}{1} = \frac{7{,}640}{2}$$

$$S_s' = \frac{7{,}640 \times 1}{2} = 3{,}820 \text{ psi}$$

In comparing the relative strengths of two shafts, the torque T is a measure of the strength of any shaft. Therefore, the comparison can be made between their torques, providing that the safe unit shearing stress has not been exceeded for either shaft.

Example. A *hollow shaft* has an external radius of 1 in. and an internal radius of 0.6 in. The strength of the shaft is to be compared with the strength of a solid shaft having the *same cross-sectional area.* The allowable unit shearing stress for each shaft is 12,000 psi.

Area of hollow shaft $= \pi r_2^2 - \pi r_1^2 = \pi(1)^2 - \pi(0.6)^2 = 0.64\pi$ sq in.

Area of hollow shaft = area of solid shaft

$$0.64\pi = \pi r^2$$

$$r^2 = 0.64$$

$$r = (0.64)^{1/2} = 0.8 \text{ in. (radius of solid shaft)}$$

Hollow shaft

$$J = \frac{\pi}{2}(r_2^4 - r_1^4) = \frac{\pi}{2}[(1)^4 - (0.6)^4]$$

$$= \frac{\pi}{2}(1 - 0.1296) = 1.367 \text{ in.}^4$$

$$T_1 = \frac{S_s J}{r}$$

$$= \frac{12{,}000 \times 1.367}{1}$$

$$= 16{,}404 \text{ in.-lb}$$

Solid shaft

$$J = \frac{\pi r^4}{2} = \frac{\pi(0.8)^4}{2}$$

$$= \frac{\pi}{2}(0.4096)$$

$$= 0.644 \text{ in.}^4$$

$$T_2 = \frac{S_s J}{r}$$

$$= \frac{12{,}000 \times 0.644}{0.8}$$

$$= 9{,}660 \text{ in.-lb}$$

Hence

$$\frac{T_1}{T_2} = \frac{16{,}404}{9{,}660} = 1.698$$

Thus, the hollow shaft is 1.698 times as strong as a solid shaft with an *equal cross-sectional area*, when made of the same material.

3-3. Angle of Twist in Shafts. As long as the shearing stress in a shaft does not exceed the elastic limit for that material, the angle of twist in a cylindrical shaft can be calculated. Referring to Fig. 3-6, which is Fig. 3-3 repeated for convenience, we note that the original reference line on the cross section of the shaft has rotated from the B position to the B' position, having moved through an angle Θ (called *theta*). The arc displacement BB' is the total angular deformation, and Θ is the angle of twist. The arc length is the product of the radius r and the angle of twist Θ, or the arc $= r\Theta = e_s$.

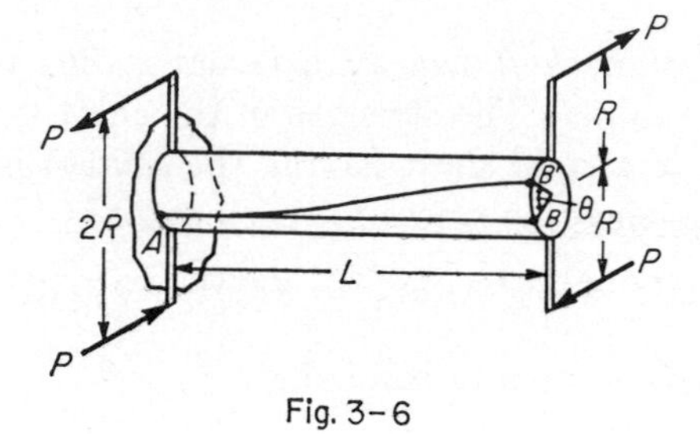

Fig. 3-6

From the statement of Hooke's law, stress is proportional to strain. Hence, for shearing conditions

$$E_s = \frac{S_s}{\varepsilon_s}$$

However,

$$\varepsilon_s = \frac{r\Theta}{L}$$

where L is the length of the shaft AB. Then

$$E_s = \frac{S_s}{r\Theta/L} = \frac{S_s L}{r\Theta}$$

where E_s = modulus of elasticity in *shear*
S_s = shearing stress at outside surface of shaft
L = length of shaft in *inches*
Θ = angle of twist in *radians*
r = outside radius in *inches*

The angle of twist can be evaluated in terms of the torque T as follows:

$$S_s = \frac{Tr}{J} \qquad \text{from} \qquad T = \frac{S_s J}{r}$$

$$S_s = \frac{E_s r \Theta}{L} \qquad \text{from above}$$

Equating the values of S_s,

$$\frac{Tr}{J} = \frac{E_s r \Theta}{L}$$

Solving for Θ,

$$\Theta = \frac{TrL}{E_s r J} = \frac{TL}{E_s J}$$

Example. A shaft which is 6 ft long and 2 in. in diameter is twisted by a torque of 18,000 in.-lb. The shaft material is steel. What is the total angle of twist in the shaft in degrees?

$$L = 6 \times 12 = 72 \text{ in.}$$

$$J = \frac{\pi}{2}(1)^4 = 1.571 \text{ in.}^4$$

$$E_s = 12{,}000{,}000 \text{ psi}$$

Then
$$\Theta = \frac{18{,}000 \times 72}{12{,}000{,}000 \times 1.571} = 0.0687 \text{ rad}$$

Converting the radians to degrees, there being 57.3° in 1 rad,

$$\Theta = 0.0687 \times 57.3 = 3.94°$$

Example. A steel shaft is 3 ft long with a diameter of 1 in. If the allowable shearing stress is limited to 10,000 psi and the angle of twist shall not exceed 1°, is the shaft safe? In an example where all the values are given, it is best to solve for the stress and compare the calculated value with the allowable stress.

$$L = 3 \times 12 = 36 \text{ in.}$$

$$\Theta = \frac{1}{57.3} = 0.0174 \text{ rad}$$

Then
$$S_s = \frac{E_s r \Theta}{L} = \frac{12{,}000{,}000 \times \frac{1}{2} \times 0.0174}{36} = 2{,}900 \text{ psi}$$

The calculated stress of 2,900 psi is less than the allowable stress of 10,000 psi; hence, the shaft is safe.

A two-piece shaft (AB and BC) is twisted as shown in Fig. 3-7. T_A = 1,800 ft-lb; T_B = 3,000 ft-lb; T_C = 1,200 ft-lb. The diameter of both pieces is 2 in. and $E_s = 12 \times 10^6$ psi. What is the maximum angle of twist in the pieces, and what is the relative angle of twist from end to end? What is the shearing stress in each part?

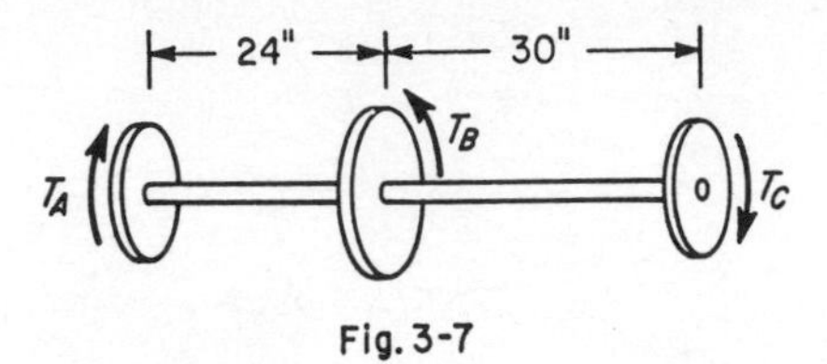

Fig. 3-7

For part AB, which is in a condition of static equilibrium with T_A = 1,800 ft-lb clockwise and a part of T_B = 1,800 ft-lb counterclockwise, T_A = 1,800 ft-lb = 21,600 in.-lb.

$$J = \frac{\pi r^4}{2} = \frac{\pi}{2} = 1.57 \text{ in.}^4$$

and
$$\theta = \frac{TL}{E_s J} = \frac{21{,}600 \times 24}{12 \times 10^6 \times 1.57} = 0.0263 \text{ rad}$$

For part BC, which is in equilibrium with a part of T_B = 1,200 ft-lb counterclockwise and T_C = 1,200 ft-lb clockwise, T = 1,200 ft-lb = 14,400 in.-lb. And

$$\theta = \frac{TL}{E_s J} = \frac{14{,}400 \times 30}{12 \times 10^6 \times 1.57} = 0.0219 \text{ rad}$$

Then, the relative angle of twist from end to end = 0.0044 rad. The shearing stress in AB is

$$S_s = \frac{21{,}600 \times 1}{1.57} = 13{,}800 \text{ psi}$$

and in BC is

$$S_s = \frac{14{,}400 \times 1}{1.57} = 9{,}170 \text{ psi}$$

3-4. Strain Energy. When a material part is subjected to an energy source such as a suddenly applied force, the *energy* which the part must absorb is termed *strain energy* (see Art. 1-11). There are sources of energy which are due to suddenly applied torques and which require that the material part be capable of absorbing the energies elastically.

Many engineering materials when subjected to a torque have a linear torque–angle of twist relation within the elastic range as illustrated

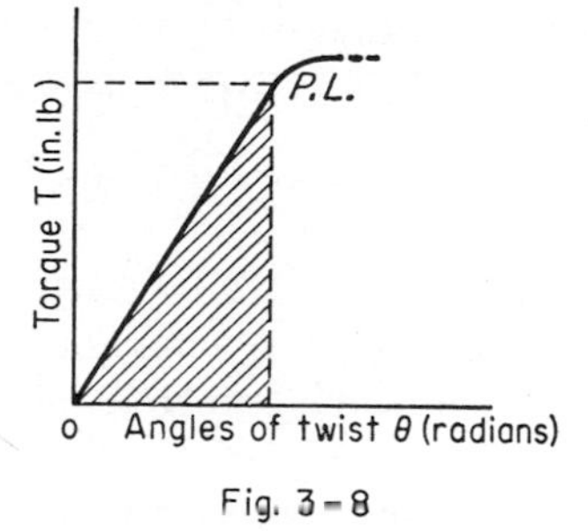

Fig. 3-8

in Fig. 3-8. Then, the total elastic energy capacity is equivalent to the shaded area (to the proportional limit) or

$$U_t = \tfrac{1}{2}T\Theta \qquad \text{in.-lb}$$

It is customary to deal with the energy capacity as a unit condition, hence

$$U = \frac{U_t}{\text{volume}} = \frac{T\Theta}{2AL} \qquad \text{in.-lb/cu in.}$$

From

$$E_s = \frac{S_sL}{r\Theta} \qquad \Theta = \frac{S_sL}{rE_s}$$

using

$$T = \frac{S_sJ}{r} \qquad \text{and} \qquad A = \pi r^2 \qquad \text{and} \qquad J = \frac{\pi r^4}{2}$$

$$U = \frac{S_s \times \pi r^4 \times S_s \times L}{2 \times r \times 2 \times r \times E_s \times \pi r^2 \times L}$$

$$U = \frac{S_s^{\,2}}{4E_s} \qquad \text{in.-lb/cu in.}$$

Good design practice dictates that the maximum allowable stress should be less than the proportional limit stress; hence only the energy equivalent to a small triangular area next to the origin should be absorbed.

An example of energy capacity or strain energy is illustrated by the following: A 2024T aluminum tube with a net cross-sectional area of 1.3 sq in. and a length of 3 ft has a maximum allowable shear strength of 8,000 psi. What is its total strain energy capacity?

From Table 1-1 $E_s = 4 \times 10^6$ psi

Then $$U = \frac{S_s^2}{4E_s} = \frac{(8,000)^2}{4 \times 4 \times 10^6} = 4 \text{ in.-lb/cu in.}$$

and $$U_t = 4 \times \text{volume} = 4 \times 1.3 \times 36 = 187.2 \text{ in.-lb}$$

Unit Energy

3-5. Power Transmission by Shafts. One of the principal functions of a shaft is to transmit power. A common example of this is the drive shaft of an automobile which transmits power from the transmission to the differential.

In order to develop a suitable expression for the power capacity of a shaft, it is necessary to define the term power. *Power* is the time rate of doing work. Considering the couple at the left end of Fig. 3-2, and understanding that the P_1 forces are constant, then the work done in one revolution by these forces will be equal to the total work done by each force or

$$U_t = P_1 \times 2\pi R + P_1 \times 2\pi R$$

$$= 2\pi \times 2P_1R$$

but $$2P_1R = T$$

and $$U_t = 2\pi T \qquad \text{per revolution of the shaft}$$

If the shaft is rotating at N revolutions per *minute*, then

$$\text{Power} = 2\pi TN \qquad \text{in torque units per minute}$$

There are 33,000 ft-lb per minute or 396,000 in.-lb per minute in 1 horsepower (abbreviated hp). It follows that when we use the correct units for torque (inch-pounds)

$$\text{Hp} = \frac{2\pi NT}{396,000}$$

In the second example in Art. 3-2, if the shaft is rotating at 33 rpm, what horsepower is the shaft transmitting?

$$T = 90,000 \text{ in.-lb}$$

From the preceding equation

$$\text{Hp} = \frac{2\pi 33 \times 90{,}000}{396{,}000} = 47.1$$

Using the second example in Art. 3-3 as the source of the design data, what is the maximum speed of rotation of the shaft when the available horsepower is 6.28? From the given data

$$T = \frac{E_s \Theta J}{L} = \frac{12 \times 10^6 \times 0.0174 \times \pi}{36 \times 32} = 569 \text{ in.-lb}$$

Then

$$N = \frac{\text{hp} \times 396{,}000}{2\pi T} = \frac{6.28 \times 396{,}000}{2\pi \times 569}$$

and

$$N = \frac{396{,}000}{569} = 696 \text{ rpm}$$

3-6. Shaft Couplings. The pulley in Fig. 3-9 has a couple applied to it with a torque value of $2P_1R$. In keeping with the principle that in the design of material parts the assembly is in a state of equilibrium, the applied torque is balanced by a P_2r torque where one of the P_2 forces acts at the surface of the shaft A. The other force P_2 acts at the center of the shaft. Then

$$2P_1R = P_2r$$

and P_2 becomes solvable when the dimensions and P_1 are known. This is another application of the principle where the external torque is equal, and opposite, to the internal torque.

In Fig. 3-9, let the following dimensions and forces apply:

$$R = 10 \text{ in.} \qquad r = 1 \text{ in.} \qquad P_1 = 300 \text{ lb}$$

The shaft A is keyed to the pulley by a key which is ¼ in. wide and 3 in. long. What is the unit shearing stress in the key? Solving for P_2,

$$P_2 \times 1 = 2 \times 300 \times 10 \qquad \text{and} \qquad P_2 = 6{,}000 \text{ lb}$$

Knowing that

$$P_s = S_sA \qquad \text{then} \qquad S_s = \frac{P_s}{A} = \frac{P_2}{A}$$

where A is the shaded cross section (Fig. 3-10) of the key relative to the shaft. There is a slight approximation in the calculation of this area, due to the arc of the circular surface of the shaft, which can be neglected. Then

$$S_s = \frac{6{,}000}{\frac{1}{4} \times 3} = 8{,}000 \text{ psi}$$

When the ends of the two shafts are to be connected by a flanged coupling, the twisting moment (torque) is transmitted from one shaft to the other. This twisting moment develops shearing stresses in the bolts. In Fig. 3-11, the side and cross-sectional views of the coupling

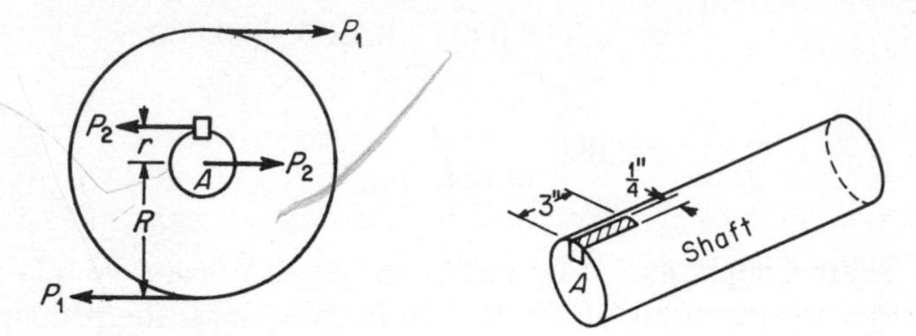

Fig. 3-9

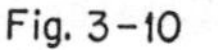

Fig. 3-10

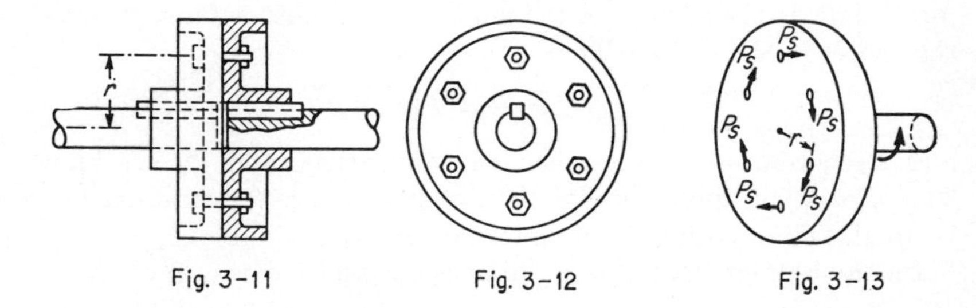

Fig. 3-11 Fig. 3-12 Fig. 3-13

are shown; in Fig. 3-12, the end view is shown; and in Fig. 3-13 a pictorial view is shown. The radius of the circle on which the bolts are placed is r. If we let P_s (see Fig. 3-13) equal the shearing strength of each bolt in pounds, the total shearing strength of all the bolts will be $P_s \times n$ where n is the number of bolts in the coupling. As

$$P_s = A_s S_s = \pi \frac{d^2}{4} S_s \qquad \text{for one bolt}$$

then the total

$$P_s = n \frac{\pi d^2}{4} S_s$$

The resisting moment of each bolt is the moment of the shearing strength about the center of the shaft, which is

$$P_s r$$

and the resisting moment of the shearing strength of all the bolts is

$$nP_s r = n \frac{\pi d^2}{4} S_s r$$

The external torque T, which is to be transmitted by means of the bolts in the coupling, will be equal to the resisting moment. Hence

$$T = n \frac{\pi d^2}{4} S_s r$$

Example. A flanged coupling is to transmit a torque of 8,000 ft-lb. The allowable shearing stress in the bolts is 10,000 psi, and the bolts are set in a circle with a 5-in. radius. How many ¾-in. bolts are required?

$$T = n \frac{\pi d^2}{4} S_s r$$

$$8{,}000 \times 12 = n \frac{\pi (3/4)^2}{4} \times 10{,}000 \times 5$$

Note. The torque must be changed to inch-pounds by multiplying by 12 in. per ft. Then

$$96{,}000 = n(0.442) \times 10{,}000 \times 5$$

(using the table in the Appendix for the total or gross area of a ¾-in. bolt), from which

$$n = \frac{96{,}000}{0.442 \times 10{,}000 \times 5} = 4.35 \text{ bolts}$$

Therefore, five bolts are required, as it is obviously impossible to use a partial bolt.

Example. A flanged coupling has six ¾-in. bolts set on a 10-in.-diameter circle. The allowable unit shearing stress in the bolts is 10,000 psi. If the shafts are rotating at 150 rpm, what horsepower can be transmitted by the coupling?

$$T = n\frac{\pi d^2}{4}S_s r$$

$$= 6 \times 0.442 \times 10{,}000 \times 5$$

$$= 132{,}600 \text{ in.-lb} = \frac{132{,}600}{12} = 11{,}050 \text{ ft-lb}$$

Then $$\text{Hp} = \frac{2\pi NT}{33{,}000}$$

$$= \frac{2\pi \times 150 \times 11{,}050}{33{,}000} = 316$$

Example. A 4-in.-diameter shaft is keyed to a pulley. The key is 6 in. long and 0.3 in. wide. When the allowable unit shearing stress in the key is 12,500 psi and the shaft is rotating at 200 rpm, what horsepower can be transmitted?

The total shearing strength of the key is

$$P_s = A_s S_s = 6 \text{ in.} \times 0.3 \text{ in.} \times 12{,}500 \text{ psi}$$

$$= 22{,}500 \text{ lb}$$

The resisting moment of the shearing strength equals the external torque

$$T = P_s r$$

$$= 22{,}500 \times 2 = 45{,}000 \text{ in.-lb}$$

and $$T = \frac{45{,}000}{12} = 3{,}750 \text{ ft-lb}$$

Then $$\text{Hp} = \frac{2\pi NT}{33{,}000}$$

$$= \frac{2\pi \times 200 \times 3{,}750}{33{,}000} = \frac{1{,}500{,}000\pi}{33{,}000}$$

$$= 142.8 \text{ which can be transmitted safely}$$

It is feasible to connect two shafts by means of (fillet) welds. An example of this is clearly shown in Fig. 3-14, where two shafts with integral flanges are butted together and joined by a fillet weld.

There are two possible methods of analysis for the torque capacity of the connection. The first of these follows the method which is used for bolted flanged connections. Letting the shear strength *per inch*

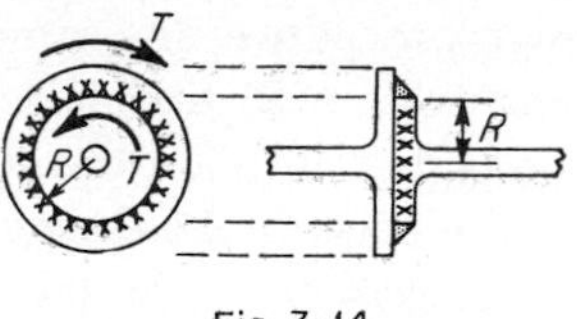

Fig. 3-14

of the fillet weld be represented by $S_s t$, then the total shearing capacity of the complete circle of welding is

$$P_s = 2\pi R(S_s t)$$

and the moment of P_s about the central axis is

$$T = P_s R = 2\pi R^2(S_s t)$$

In Fig. 3-14, with $R = 2$ in., and a 3/8-inch fillet weld with a shearing stress value of 3,600 lb per in., then

$$P_s = 2\pi R(3{,}600)$$

$$= 2\pi R(3{,}600) = 14{,}400\pi \text{ lb}$$

and $$T = P_s R = 14{,}400\pi \times 2 = 28{,}800\pi \text{ in.-lb}$$

The alternative method for determining the torque capacity makes use of the general torque equation

$$T = \frac{S_s J}{r}$$

In applying this equation, it is necessary to evaluate the polar moment of inertia of the weld, as an equivalent area, about the central axis, which gives

$$J = 2\pi R^3 t$$

and $$r = R$$

Thus $$T = \frac{S_s \times 2\pi R^3 t}{R}$$

$$= 2\pi R^2 S_s t \qquad \text{as above}$$

Note. The use of the radius R of the smaller flange as the design radius may be questioned when the throat area of a fillet weld is critical. The precise radius to the geometrical center of the throat area is the exact dimension. However, regardless of the percent approximation of the design procedure, the same idea is used in axially loaded welded connections (Chap. 2) and is followed in the analysis here. The approximation is definitely on the side of safety and serves to increase the condition of safety of the design.

Continuing with the numerical example, let it be required to determine the minimum radii of the shafts being connected, the shaft material having a design shearing stress of 8,000 psi. From

$$T = \frac{S_s J}{r}$$

$$= \frac{8{,}000\pi r^4}{r(2)} = 4{,}000\pi r^3 \text{ in.-lb}$$

Equating the torques,

$$28{,}800\pi = 4{,}000\pi r^3$$

$$r^3 = 7.2 \qquad \text{and} \qquad r = 1.93 \text{ in.}$$

Problems

3-1. A 2-in.-diameter solid steel shaft is to transmit 2,000 ft-lb of torque. What is the maximum shearing stress in the shaft?

3-2. If the shaft in Prob. 3-1 is 45 in. long, what is the angle of twist in radians?

3-3. An aluminum-alloy shaft 40 in. long is limited to an angle of twist of 2½°. The allowable shearing stress shall not exceed 11,000 psi. What is the required diameter of the shaft?

3-4. A hollow 2024T aluminum-alloy shaft with 4-in. and 2-in. diameters is directly connected to a source which supplies a torque of 10,000 ft-lb. (*a*) What is the maximum shearing stress which is developed? (*b*) What is the shearing stress at the inner surface of the tube?

3-5. Figure 3-15 shows a two-part shaft which is being used to transmit torques T_E and T_F. The shaft is solid steel, with diameters of 2 in. and 1 in., and is rigidly attached to the wall at D. What is the angle of twist in degrees at the free end F of the shaft?

3-6. What is the maximum shearing stress in the steel shafting of Prob. 3-5?

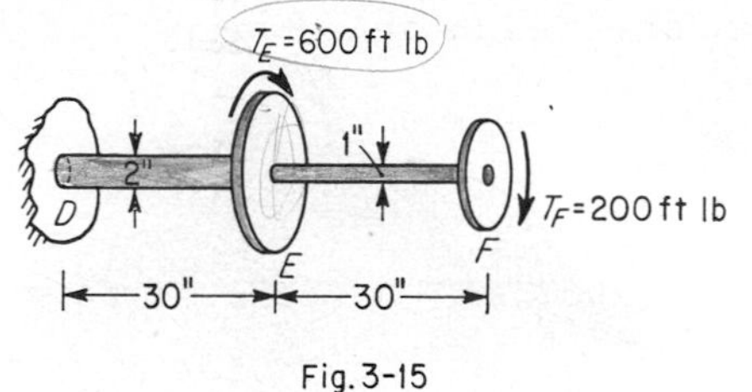

Fig. 3-15

3-7. A hollow shaft is twisted by means of a 3,000 ft-lb torque. The external and internal diameters of the shaft are, repectively, 4 in. and 3 in. The material is steel, and the angle of twist is limited to 1.5°. What is the maximum length of shaft which can be used under this specification?

3-8. A 1-in.-diameter and 40-in.-long 2024T aluminum shaft is placed in a vertical position shown in Fig. 3-16. A 20-in.-diameter plate is attached to the upper end. What arc displacement AA' results from the application of a torque T which is limited by a maximum shear stress of 10,000 psi? What is the magnitude of the limiting torque?

3-9. The shaft, shown in Fig. 3-16, is required to absorb a unit energy of 2.5 in.-lb per cu in. What maximum shearing stress is developed in the shaft?

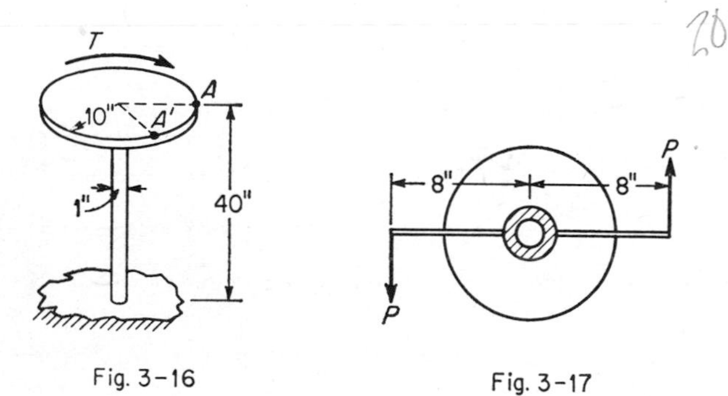

Fig. 3-16

Fig. 3-17

3-10. What is the total angle of twist (radians) in the shaft of Fig. 3-16 when used under the conditions stated in Prob. 3-9?

3-11. A torque is suddenly applied to a tube of 2024T aluminum alloy by means of a couple as shown in Fig. 3-17. The radii of the tube are 1 in. and ½ in. If the couple produces a strain energy of 4 in.-lb per cu in. in the material, what is the magnitude of each force P of the couple?

3-12. A two-part shaft has the dimensions shown in Fig. 3-18. $T_D = 6{,}000$ in.-lb; $T_E = 4{,}000$ in.-lb; $T_F = 2{,}000$ in.-lb. What is the total strain energy absorbed by the 40-in. piece of 2024T aluminum alloy?

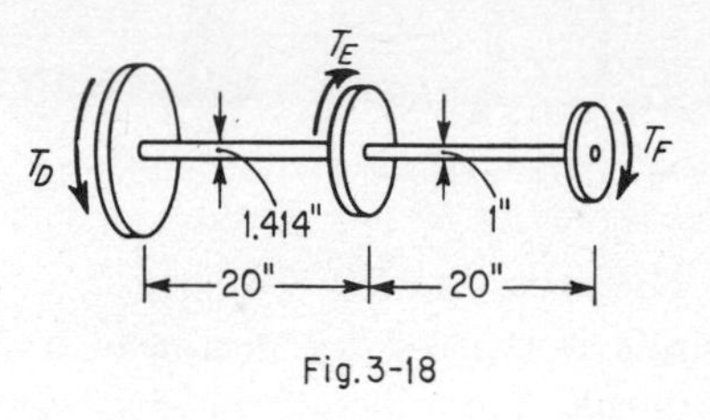

Fig. 3-18

3-13. A two-shaft assembly is shown in Fig. 3-19 in diagrammatic form. Bearings, which are not shown, keep the shafts in their respective positions but do not prevent the twisting of the shafts. There is no slipping between the wheels B and C. If a torque of 2,400 in.-lb is applied at A, through what angle will wheel C turn assuming that wheel D is fixed, and all twisting takes place with respect to it? $E = 4 \times 10^6$ psi.

3-14. Through what angle does wheel A turn relative to wheel B? (Prob. 3-13.)

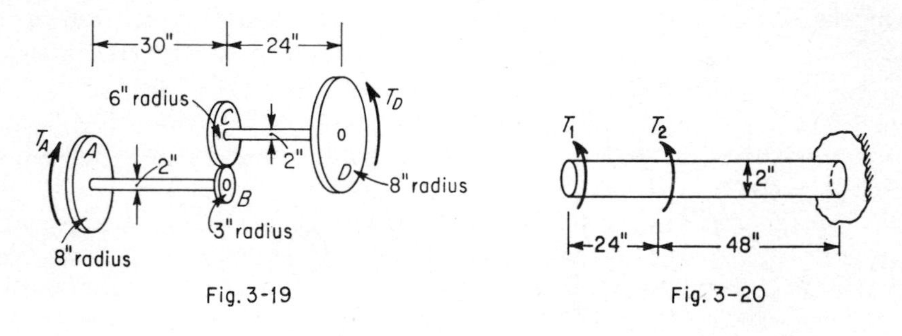

Fig. 3-19

Fig. 3-20

3-15. Through what angle does wheel A turn in degrees? (Probs. 3-13 and 3-14.)

3-16. The 2-in.-diameter steel shaft in Fig. 3-20 is limited to a maximum stress of 12,000 psi and a total angle of twist of 0.06 rad. Determine the values of T_1 and T_2.

3-17. A 3-in.-diameter solid steel shaft is used to transmit 24,000 in.-lb of torque at 99 rpm. What is the maximum unit shearing stress in the shaft?

3-18. In Prob. 3-17, how many horsepower does the shaft transmit?

3-19. An aluminum-alloy shaft 40 in. long is limited to an angle of twist of 3°. The allowable shearing stress shall not exceed 12,000 psi. What is the required diameter of the shaft?

3-20. If the shaft of Prob. 3-19 is rotating at 6.6 rpm, what horsepower is being transmitted?

3-21. A hollow shaft with outside and inside diameters of 18 in. and 12 in. respectively, transmits 10,000 hp at 99 rpm. What is the maximum shearing stress in the shaft?

3-22. In the shaft of Prob. 3-21, what is the minimum shearing stress in the shaft?

3-23. A flanged coupling is used to connect two shafts, the left and right shafts (see Fig. 3-11) being 3 in. and 2 in. in diameter, respectively. The key, keying the left shaft to its flange, is 1½ in. long and 0.24 in. wide. The key, keying the right shaft to its flange, is 1½ in. long and 0.30 in. wide. The key material has an allowable shearing strength of 12,000 psi. What is the limiting torque which can be transmitted by the assembly, as determined by the capacity of the keys?

3-24. When the assembly described in Prob. 3-23 rotates at 1,200 rpm, how many horsepower can be transmitted?

3-25. A pulley is keyed to a shaft, which in turn is connected to another shaft by a flanged coupling. The assembly transmits 40 hp at 165 rpm. What torque does the assembly transmit?

3-26. Using the data from Prob. 3-25, how many ½-in.-diameter bolts arranged in a 4-in.-radius circle are required to connect the flanges if the bolt material has an allowable shear strength of 6,000 psi?

3-27. When the two parts of the shaft assembly shown in Fig. 3-21 are welded together by a ¼-in. fillet weld, what is the maximum shearing stress in the shaft material?

3-28. Using the data in Prob. 3-27, what is the total angle of twist in the 60-in. length of steel (in degrees)?

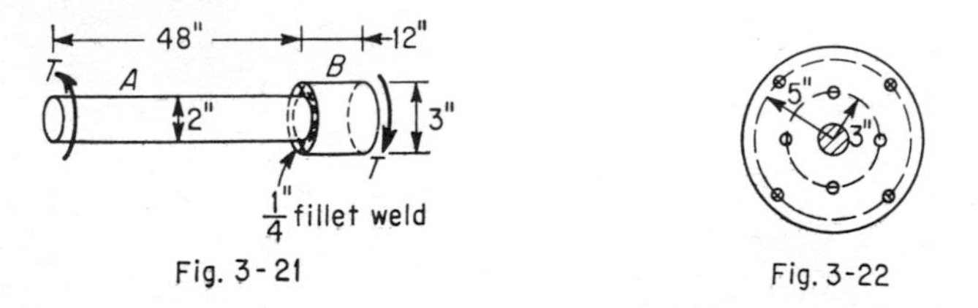

Fig. 3-21

Fig. 3-22

3-29. At what speed must the shaft of Prob. 3-27 rotate to deliver 40 hp?

3-30. A flanged coupling as shown in Fig. 3-22 has eight ⅜-in.-diameter bolts arranged as shown. If the coupling is to transmit

2,500 ft-lb of torque, determine the maximum shearing stress in the bolts.

3-31. Figure 3-23 shows a flanged coupling with four ½-in.-diameter bolts with an allowable shearing strength of 12,000 psi. The left is welded to its flange by a continuous ⅜-in. fillet weld. The right shaft is integral with its flange. Determine the torque capacity of the assembly as limited by the weld or bolts.

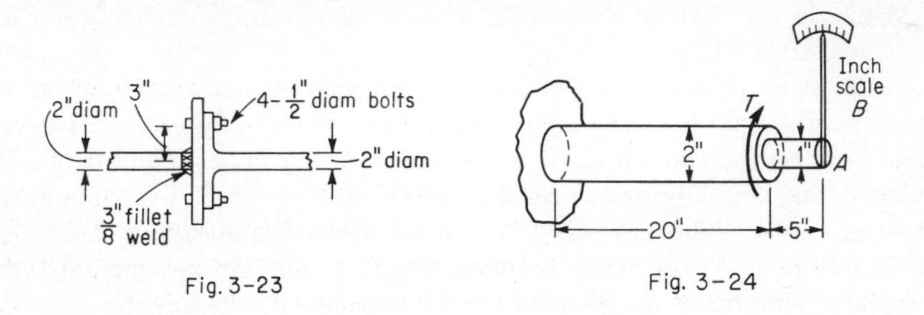

Fig. 3-23 Fig. 3-24

3-32. In Prob. 3-31, determine the maximum shearing stress in either shaft.

3-33. A two-part shaft shown in Fig. 3-24 is twisted by a torque $T = 6{,}000\pi$ in.-lb. A weightless pointer is rigidly attached to the shaft at A and is 12 in. long, measured from the central axis. The scale B is in the plane of the end A of the shaft. If the material is 2024T aluminum alloy, what is the arc reading (in.) on the scale B?

CHAPTER 4

BEAMS—SHEAR AND MOMENT DIAGRAMS

4-1. Beams. In Chap. 1, only those material parts which were subjected to axial forces and their resisting tensile or compressive stresses were analyzed. In Chap. 2, those material parts which were acted upon by forces which cause internal resisting shearing stresses were discussed. In Chap. 3, material parts which were twisted by torques or couples were analyzed, and the resisting shearing stresses could be calculated. In this chapter, those material parts which are acted upon by transverse loads will be discussed.

A *beam* is a piece of material which is subjected to forces which act transversely (perpendicular) to the long axis of the piece. This text will consider only those cases where:

1. All applied forces or loads act in a single plane which corresponds to one of two principal axes of the cross section of the beam.
2. All beams are prismatic unless otherwise stated.
3. All beams are straight and are assumed to be in the horizontal position (although they may be placed vertically or diagonally).
4. All beams are statically determinate, and therefore the equations of static equilibrium apply to them.
5. All beams are presumed to be laterally braced to prevent overturning and/or twisting.

There are numerous engineering examples of beams such as floor joist, structural beams in buildings and bridges, automobile axles, and levers.

4-2. Beam Supports. The availability of the basic equations of equilibrium

$$\Sigma F_x = 0 \qquad \Sigma F_y = 0 \qquad \text{and} \qquad \Sigma M = 0$$

limits to three the number of unknown reactions which can be solved in a statically determinate beam. Two of these may be components of a reaction with an unknown direction. Two engineering diagrams are used to denote specific types of reactions. In each case, the left or *a* diagram is equivalent to the right or *b* diagram.

Figure 4-1*a* is the engineering equivalent of the roller reaction pinned to the beam shown in Fig. 4-1*b*. Figure 4-3 shows that the reaction for (*a*) must be perpendicular to the surface on which the rollers could move, since the reaction offers no resistance (force) to motion along this surface. The roller reaction also implies that its force can either *push* or *pull* on the pin as is necessary for equilibrium.

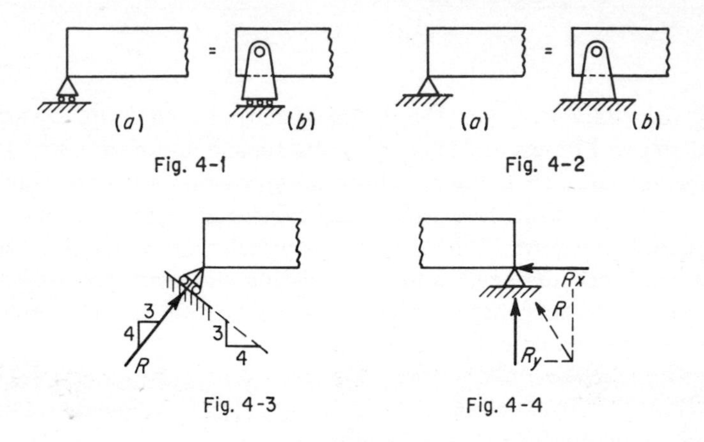

Fig. 4-1 Fig. 4-2

Fig. 4-3 Fig. 4-4

Figure 4-2*a* is the engineering equivalent of a support which is fixed to the surface on which it is resting. Figure 4-2*b* shows the actual configuration of the support, while Fig. 4-4 gives the components of the reaction (with the reaction indicated by the dashed line). Figure 4-2*b* shows a fixed pin through the beam, which allows the beam to sag (deflect) without restraint. This is also true of the pin action in Fig. 4-1*b*.

4-3. Types of Beams. There are three general types of statically determinate beams:

1. Simple beam
2. Overhanging beam
3. Cantilever beam

In the first and second categories the beams are supported at two points, and in the third category the beam is supported at one end. These are illustrated diagrammatically in Figs. 4-5 to 4-7. Note that the engineering diagrams represent the beams by single straight lines.

4-4. Loads on Beams. Loads which act (usually downward) on beams may be concentrated, uniform, or nonuniform. A *concentrated load* is one which is considered as being applied at a single point on the beam. Concentrated loads are usually expressed in pounds or kips. A *uniform load* is one which is spread uniformly over a length of the beam. Uniform loads are expressed, generally, in pounds per foot or kips per foot. *Nonuniform loads* are similar to uniform loads except that they vary in magnitude per foot. Only those nonuniform loads

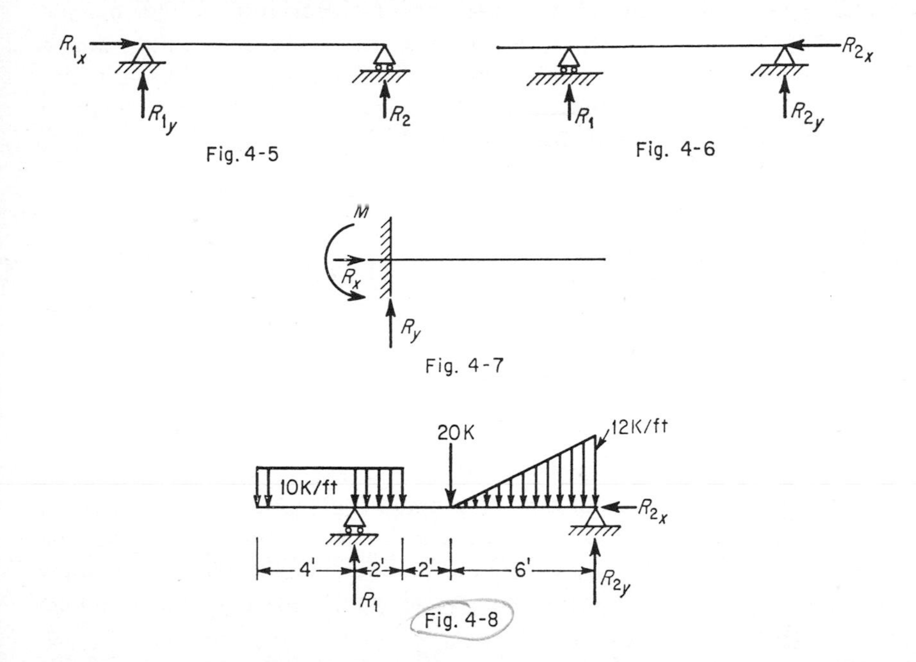

that have a linear variation in magnitude will be considered in this text. Nonuniform loads are expressed by their end ordinate values in pounds or kips per foot (of beam length). Figure 4-8 illustrates the three types of loading on an overhanging beam.

The uniform loading extends over the left 6 ft of the beam; a concentrated load is applied 8 ft from the left end of the beam; a nonuniform load with a right end ordinate at the rate of 12 kips per ft extends over the right 6 ft of the beam.

4-5. Beam Reactions. The first step in the design of a material part to be used as beam requires that the beam must be in a condition of static equilibrium. A free-body diagram of all the forces which act on the beam *should be sketched.*

For the *sole purpose* of solving the reactions, each uniform load should be replaced by its resultant forces acting at the middle of the length of the uniform load. Note that the total uniform load in Fig. 4-8 is equivalent to the area of the representing rectangle with an ordinate value of 10 kips per ft and a base length of 6 ft. The resultant force equals 60 kips acting 3 ft from either end of the uniform load, or 3 ft from the left end of the beam.

Also, a nonuniform load should be replaced by its resultant force acting vertically through the centroid of the triangle. The magnitude of this load is equal to the area of the representing triangle or is equal to one-half of the base times the altitude. Numerically, the total nonuniform load is 36 kips acting 2 ft from the right end or 4 ft from the left end of this loading.

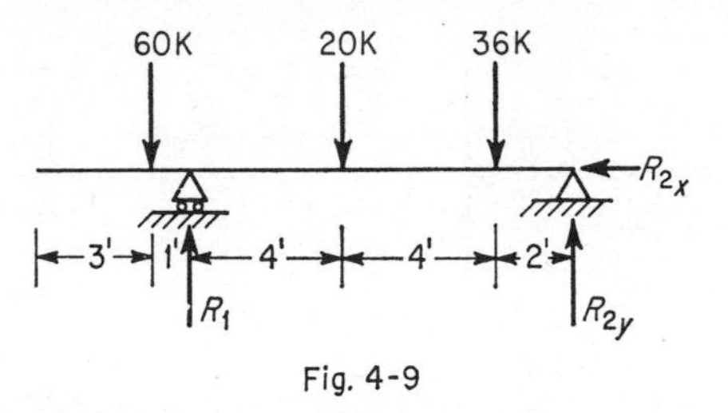

Fig. 4-9

The diagrammatic sketch of the complete free-body diagram is shown in Fig. 4-9. Using this diagram of the beam, solve the vertical reactions by $\Sigma M_{R_1} = 0$ and $\Sigma M_{R_2} = 0$. By so doing, $\Sigma F_y = 0$ is available for checking your calculations. This is important in order to ensure correct reaction values which will be used in a continuous set of calculations.

By $\Sigma M_{R_1} = 0$ and using the arbitrary clockwise convention as plus (+),

$$-10R_{2_y} - 1 \times 60 + 4 \times 20 + 8 \times 36 = 0$$

$$-10R_{2_y} + 308 = 0$$

$$R_{2_y} = 30.8 \text{ kips}$$

By $\Sigma M_{R_{y_2}} = 0$ and using the same sign convention,

$$+10R_1 - 11 \times 60 - 20 \times 6 - 36 \times 2 = 0$$

$$+10R_1 - 852 = 0$$

$$R_1 = 85.2 \text{ kips}$$

By $\Sigma F_y = 0$ to check,

$$+R_1 + R_{2_y} - 60 - 20 - 36 = 0$$
$$30.8 + 85.2 = 116$$
$$116 = 116$$

A $\Sigma F_x = 0$ equation shows that $R_{x_2} = 0$.

The calculations of the reactions for a cantilever-type beam are slightly different from those which were just illustrated. This is due entirely to the situation where the structure is supported at one end by a fixed support which must provide the total external reaction system. In Fig. 4-7, the reaction system must supply a vertical force and a moment M (equivalent to a couple), so that the system of forces acting on the beam is in a condition of static equilibrium.

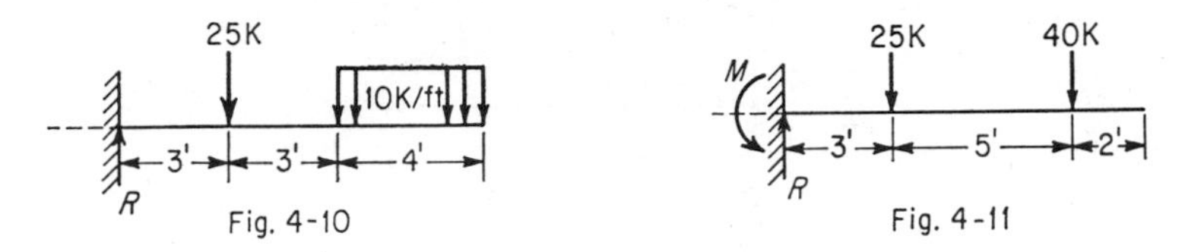

Fig. 4-10 Fig. 4-11

Figure 4-10 shows a cantilever beam loaded with a partial uniform load and a concentrated load. The free-body diagram of this beam is sketched in Fig. 4-11 and is used for the *sole purpose* of solving the reactions.

By $\Sigma F_y = 0$,

$$+R - 25 - 40 = 0$$
$$R = 65 \text{ kips}$$

By $\Sigma M_R = 0$ and using the clockwise convention (+),

$$+3 \times 25 + 8 \times 40 - M = 0$$
$$+395 - M = 0$$
$$M = +395 \text{ ft-kips}$$

Note. The plus sign is an *algebraic indication* that the turning direction assumed for the moment is correct. If the answer for M had been minus, the minus sign would have indicated an incorrect turning-direction assumption for M.

To check the value of M, write a $\Sigma M = 0$ equation about some other axis (point) such as the right end of the beam. Thus ΣM about the right end = 0 and is

$$-2 \times 40 - 7 \times 25 + 10 \times 65 - 395 = 0$$
$$-650 + 650 = 0$$

In the second preceding equation, note that the moment, with its correct sign convention, must be included. The moment is the same as the moment value of a couple which has a constant turning magnitude about any axis (point) perpendicular to the plane containing the couple.

4-6. Shear at a Section of a Beam. When a beam is in a condition of static equilibrium, any part of the beam is likewise in a condition of equilibrium. Referring to Fig. 4-8, the free-body diagram of the left 6 ft of the beam is shown in Fig. 4-12. (Note that the beam is now

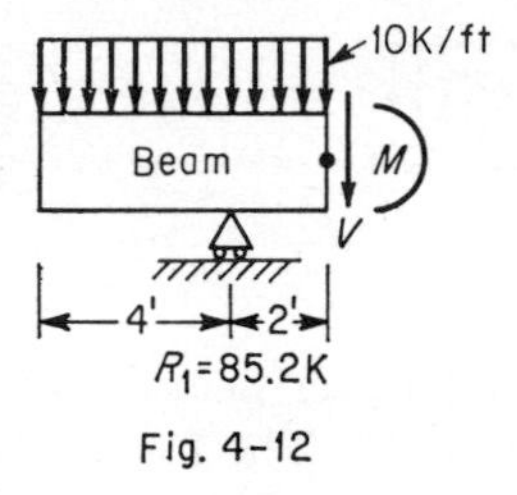

Fig. 4-12

shown as a rectangular bar.) The remaining portion of the beam is not considered. The three equations of static equilibrium must be satisfied. In this article, only $\Sigma F_x = 0$ and $\Sigma F_y = 0$ will be evaluated.

At the right end of this part of the beam, the internal *force system* exerted on this part by the discarded portion of the beam is shown. This force system logically includes a vertical force V and a moment M. It now becomes necessary to determine the value of V. Applying $\Sigma F_y = 0$, with the positive convention upward and the sense of V assumed to be downward,

$$\Sigma F_y = 0 = -6'(10 \text{ kips/ft}) + 85.2 \text{ kips} - V$$
$$0 = -60 + 85.2 - V$$
$$V = +25.2 \text{ kips}$$

The plus sign indicates that the *sense* of V for equilibrium was assumed correctly.

The *vertical shear* at any section of a beam is the *vector sum* of the vertical forces to the left of the section. The term *section* is the name given to the cross section of the beam where V is being calculated. This value of the vertical shear is always equal in magnitude to the value of V as determined for the condition of equilibrium, but its sense is opposite to that of the calculated value. This is true because the external forces must be balanced by an equal and opposite internal force, and the vertical shear V is equivalent to the resultant external force system.

The conventional sign for the vertical shear is determined thus: When the *vector sum* of the external vertical forces acting on the part of the beam to the left of the section is *upward*, the shear is *positive;* when the *vector sum* of the vertical forces to the left of the section is *downward*, the shear is *negative*. These conventions imply that a positive vertical shear indicates that the beam has a tendency to be

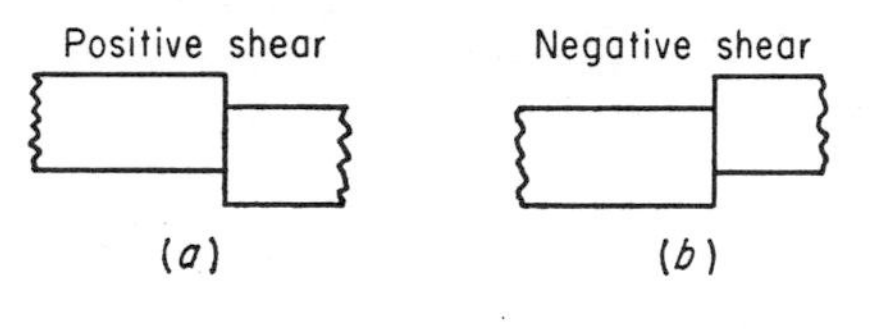

Fig. 4-13

sheared or cut at the section shown in Fig. 4-13*a*. A negative vertical shear indicates the tendency to shear as shown in Fig. 4-13*b*.

The designing engineer is interested in the maximum value of the vertical shear in the beam. For this reason, and others which will be explained in a later article, it is convenient to determine the vertical shear at many sections of the beam and plot them with respect to a horizontal zero-shear axis.

In order to plot and determine the correct shape of the area of the shear diagram, certain values become critical. Shear values should be obtained at sections which are located:

1. Just to the left *and* right of each concentrated force. Reactions are considered as concentrated forces.
2. At the beginning *and* end of each uniform and nonuniform loading.
3. At as many intermediate points within the lengths of the uniform and nonuniform loadings as are deemed necessary to show the approximate shape of the area.

Again referring to Fig. 4-8, we find the values of V are determined at the distances from the *left* end as follows:

0 ft:

$$V = 0$$

2 ft:

$$V = -2(10 \text{ kips/ft}) = -20 \text{ kips}$$

4 ft_L:

$$V = -4(10 \text{ kips/ft}) = -40 \text{ kips}$$

4 ft_R:

$$V = -4(10 \text{ kips/ft}) + 85.2 = +45.2 \text{ kips}$$

6 ft:

$$V = -6(10 \text{ kips/ft}) + 85.2 = +25.2 \text{ kips}$$

8 ft_L:

$$V = -6(10 \text{ kips/ft}) + 85.2 = +25.2 \text{ kips}$$

8 ft_R:

$$V = -6(10 \text{ kips/ft}) + 85.2 - 20 = +5.2 \text{ kips}$$

10 ft:

$$V = -6(10 \text{ kips/ft}) + 85.2 - 20 - \tfrac{1}{2}2(4 \text{ kips/ft}) = +1.2 \text{ kips}$$

12 ft:

$$V = -6(10 \text{ kips/ft}) + 85.2 - 20 - \tfrac{1}{2}4(8 \text{ kips/ft}) = -10.8 \text{ kips}$$

14 ft_L:

$$V = -6(10 \text{ kips/ft}) + 85.2 - 20 - \tfrac{1}{2}6(12 \text{ kips/ft}) = -30.8 \text{ kips}$$

14 ft_R:

$$V = -6(10 \text{ kips/ft}) + 85.2 - 20 - \tfrac{1}{2}6(12 \text{ kips/ft}) + 30.8 = 0$$

In the tabulation, the subscripts R and L refer to the sections at the immediate right and left of the specified distances from the left end of the beam.

It should be noticed that $V = 0$ at the ends of the beam. It is also possible to write a series of algebraic equations for the shear values at the various sections of the beam. An example of the equations for the vertical shear at any section between the distances of 4 and 6 ft from the left end of the beam is (letting X equal the distance from the left end of the section)

$$V = -X(10 \text{ kips/ft}) + 85.2 \text{ kips}$$

Substituting 4 ft and 6 ft for X in this equation will check the values at 4 ft_R and 6 ft as above.

The shear diagram is then sketched or plotted to scale from a horizontal zero base line directly under the sketch of the loading diagram of the beam. Positive values of shear are conventionally plotted above the zero axis. Figure 4-14 shows the plotted tabulated values with the several plotted points connected by smooth curves or straight lines. The uniform and nonuniform loads cause the shear diagram to

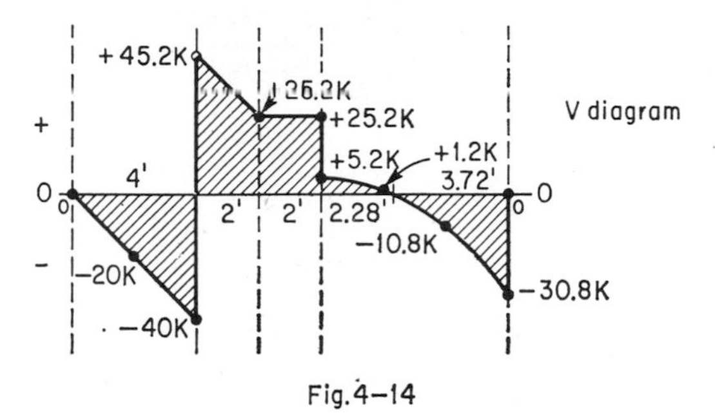

Fig. 4-14

be respectively sloped or curved. Concentrated loads cause vertical lines in the shear diagrams. A part of the beam supporting no loading causes the diagram to be a straight horizontal line.

As it will be learned later, the sections of the beam at which the shear is zero are important, and their positions along the beam should be located. Not counting the zero shear points at the ends of the beam, there are two sections of zero shear. The location, 4 ft from the left end, of one of these is obvious. The other one requires some calculations. Figure 4-15 shows the pertinent parts of Figs. 4-8 and 4-14 and will be used as a basis of locating the other section of zero shear.

The magnitude of the nonuniform loading Y at any distance X from the left end of this loading is the rate of loading

$$\frac{12 \text{ kips/ft}}{6 \text{ ft}}(X)$$

where X varies from 0 to 6 ft. Applying the general principle for determining the vertical shear at a section of the beam, but starting with the vertical shear value at 8 $ft_R = +5.2$ kips, then

$$V = 0 = +5.2 - {}^{12}\!/_{6}(X)(\tfrac{1}{2})(X)$$

$$0 = +5.2 - X^2$$

$$X = (5.2)^{\frac{1}{2}} = 2.28 \text{ ft}$$

The shear diagram for a cantilever beam is drawn in the same general manner as that in the preceding example. In Figs. 4-10 and 4-11,

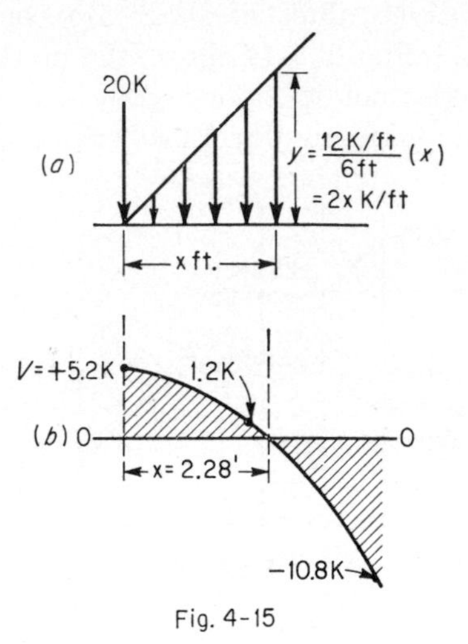

Fig. 4-15

$R = 65$ kips. Tabulating the shear values at distances from the left end as follows:

0 ft_L:	$V = 0$
0 ft_R:	$V = +65$ kips
3 ft_L:	$V = +65$ kips
3 ft_R:	$V = +65 - 25 = +40$ kips
6 ft:	$V = +65 - 25 = +40$ kips
8 ft:	$V = +65 - 25 - 20 = +20$ kips
10 ft:	$V = +65 - 25 - 40 = 0$

The plotted shear diagram is shown in Fig. 4-16. Shear diagrams should always be checked by calculating the values of the shear from end to end of the beam.

4-7. Moments and Moment Diagrams. The magnitude of the moment at any section of a beam is defined as the algebraic sum of the moments of all the forces on the left or right of the section about an axis (point) *at that section.* If we refer to Fig. 4-12, and the value of the internal moment for equilibrium is to be determined for the section 6 ft from the left end of the beam, ΣM must equal zero for this part of

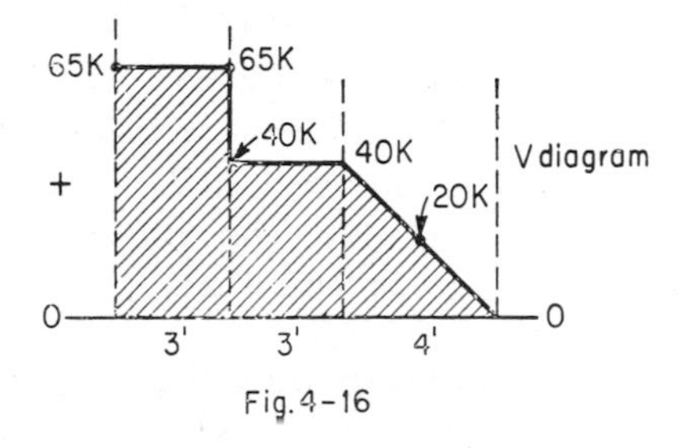

Fig. 4-16

the beam. The moment M which is shown becomes a part of the free-body diagram. Writing the moment equation about the dot on the section, and using clockwise turning as positive,

$$\Sigma M = 0 = -(6')10 \text{ kips/ft } (3') + 85.2(2) + M$$

$$0 = -180 + 170.4 + M$$

$$M = +9.6 \text{ ft-kips}$$

The plus sign indicates the correct assumption for M for equilibrium.

This value of the internal moment is always equal in magnitude to the external bending moment M but of opposite sense. This is another way of stating

$$\Sigma M_{\text{ext}} = \Sigma M_{\text{int}}$$

where the external moment is the required moment.

Bending moment is the term which is applied to the external moment at a section of the beam since the bending moments are the reason for the bending of the loaded beam.

Of the several available methods for setting the conventional sign of the moment, the following rule is the simplest and easiest to apply: *The moment produced by any upward force is positive; the moment produced by any downward force is negative.* This rule eliminates the necessity for remembering a clockwise or counterclockwise convention system when the moments are produced by concentrated forces or resultant forces from uniform or nonuniform loadings.

The designing engineer is also interested in the magnitudes and signs of the critical moments as well as the locations of the sections of these critical values. As was the case with the shear diagram, it is helpful to determine the moments at many sections of the beam and plot them with respect to a horizontal zero-moment axis. The general shape of the moment diagram is useful, so the bending moments should be calculated:

1. At each concentrated force or reaction
2. At both ends of each uniform and nonuniform loading
3. At several intermediate points within the lengths of the uniform and nonuniform loadings to determine the true shape of the area
4. At all sections at which the shear is zero

Referring to Fig. 4-8 and working with the left end (although the moments can be determined just as well when working with the right end), we find the values of M are as follows:

0 ft: $M = 0$

2 ft: $M = -(2)\ 10\ \text{kips/ft}\ (1) = -20\ \text{ft-kips}$

4 ft: $M = -(4)\ 10\ \text{kips/ft}\ (2) = -80\ \text{ft-kips}$

6 ft: $M = -(6)\ 10\ \text{kips/ft}\ (3) + (2)\ 85.2\ \text{kips} = -9.6\ \text{ft-kips}$

8 ft: $M = -(6)\ 10\ \text{kips/ft}\ (5) + (4)\ 85.2\ \text{kips} = +40.8\ \text{ft-kips}$

At the section 10 ft from the left end, 2 ft of the nonuniform loading is included. Figure 4-15*a* gives an ordinate value of 4 kips per ft and a total loading of $\frac{1}{2}(2)(4$ kips per ft$) = 4$ kips. This resultant force acts at the *centroid* of the triangular loading which is $\frac{2}{3}(2) = \frac{4}{3}$ ft from the left end of the triangle or $\frac{2}{3}$ ft from the right end of the triangle. Then at

10 ft: $$M = -(6)\ 10\ \text{kips/ft}\ (7) + (6)\ 85.2\ \text{kips} - (2)\ 20\ \text{kips} - (\tfrac{2}{3})4\ \text{kips} = +48.5\ \text{ft-kips}$$

Similarly

10.28 ft: $$M = -(6)\ 10\ \text{kips/ft}\ (7.28) + (6.28)\ 85.2\ \text{kips} - (2.28)\ 20\ \text{kips} - (2.28/3)(2.28/2)\ 4.56 = +48.7\ \text{ft-kips}$$

12 ft: $M = -(6)\ 10\ \text{kips/ft}\ (9) + (8)\ 85.2\ \text{kips} - (4)\ 20\ \text{kips} - (\tfrac{4}{3})\ 16\ \text{kips}$

$= +40.3$ ft-kips

14 ft: $M = -(6)\ 10\ \text{kips/ft}\ (11) + (10)\ 85.2\ \text{kips} - (6)\ 20\ \text{kips} - (2)\ 36\ \text{kips}$

$= 0$

Figure 4-17 shows the complete moment diagram as plotted from the preceding values.

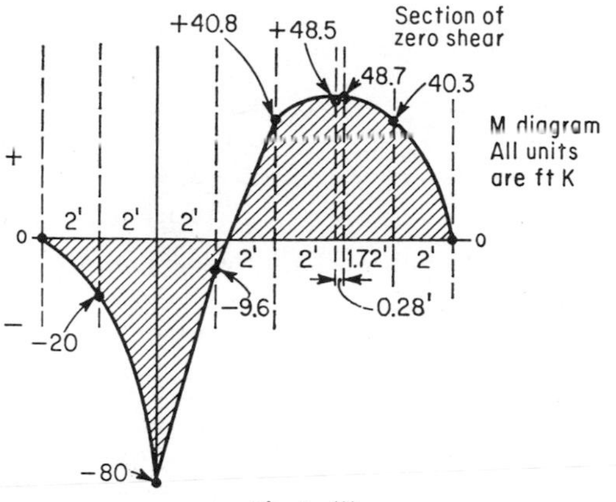

Fig. 4-17

An examination of the general shape of the moment diagram indicates that a uniform loading results in a curved diagram (actually this is a second-power mathematical curve); no loading gives a straight-line diagram; nonuniform loading gives a curved diagram (this is a third-power mathematical curve).

The direction and magnitude of the bending of a beam are of importance to the designer. This article will consider only the *direction* of the *bending.*

When the bending moment in a part of a beam is positive, the beam is curved as shown in Fig. 4-18. When the bending moment is negative, the bending is as shown in Fig. 4-19. The conventional curvature shown in the figures is exaggerated in order to show the general direction of bending, and the diagrams do not imply that the ends of the curves lie on a horizontal line. In fact, the only fixed point or points on the bending (elastic) curve occur at the reactions. All bending takes place relative to the fixed point or points. The *elastic curve* is a continuous smooth curve without abrupt changes, since it represents the curvature in a piece of material.

It follows that when the bending moment is zero, the beam is not bending; hence the section of the beam at which $M = 0$ is straight.

In Fig. 4-17, only one of three sections of zero moment need be located. There are two available methods for locating this particular section. Since the length of beam from 6 ft to 8 ft is unloaded, the moment diagram is a straight line, and the section of zero M can be

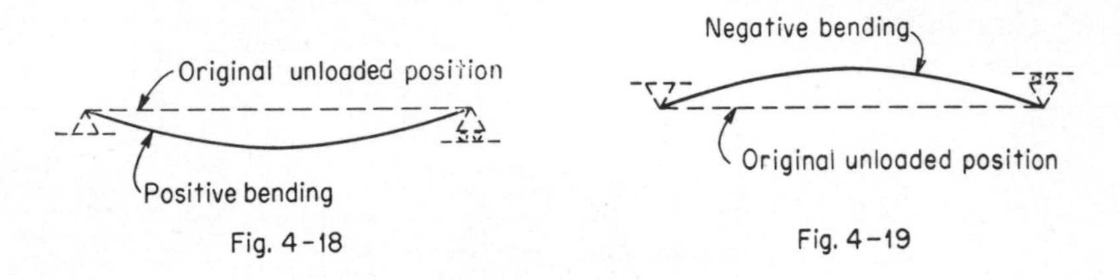

Fig. 4-18 Fig. 4-19

located by proportion. Letting X equal the distance from 6 ft to $M = 0$, and omitting the algebraic signs of the plus and minus moments at 6 ft and 8 ft, then

$$9.6:X::40.8:2 - X$$

from which $$X = 0.38 \text{ ft}$$

The alternate method which is usually required makes use of a moment equation about the section of zero moment. Again letting X equal the distance from 6 ft to $M = 0$,

$$M = 0 = -(6)\ 10 \text{ kips/ft } (3 + X) + (2 + X)\ 85.2 \text{ kips}$$

from which

$$X = 0.38 \text{ ft}$$

Having located this section of zero moment, we can sketch the bending curve. Before any loading is applied to the beam, the beam is straight and is supported by the reactions. Making use of the conventional directions of bending, Fig. 4-20 shows the approximate bending curve. Point I, the section of zero moment, can be above the original position of the beam even though it is shown below. Point I is called the *point of inflection*, or the point at which the elastic curve changes direction of bending. The elastic curve on either side of this point is tangent to the infinitesimal straight length of the curve which is shown extended as line TT.

In Figs. 4-10, 4-11, and 4-16, which are the loading and shear diagrams of a cantilever beam, the moments at sections at the wall, 7 ft, 4 ft, and 2 ft from the right end should be obtained and plotted for

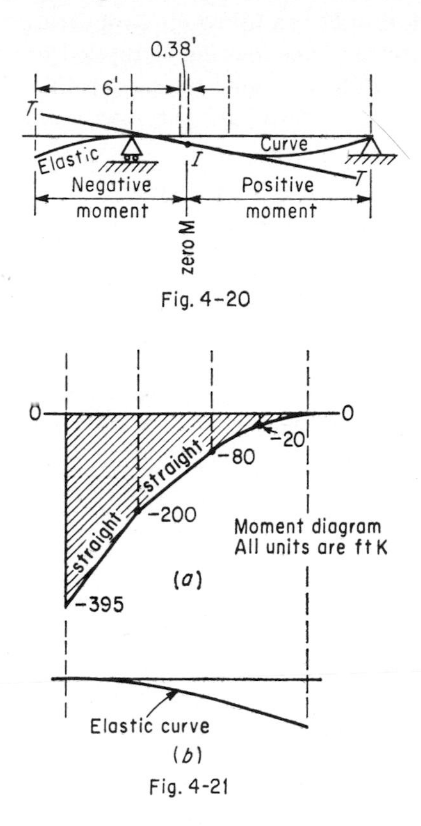

Fig. 4-20

Fig. 4-21

the moment diagram. In these calculations, the part of the beam to the right of the section will be used.

At wall: $M = -(3)\ 25\ \text{kips} - (8)\ 10\ \text{kips/ft}\ (4) = -395\ \text{ft-kips}$

7 ft: $M = -(5)\ 10\ \text{kips/ft}\ (4) = -200\ \text{ft-kips}$

4 ft: $M = -(2)\ 10\ \text{kips/ft}\ (4) = -80\ \text{ft-kips}$

2 ft: $M = -(1)\ 10\ \text{kips/ft}\ (2) = -20\ \text{ft-kips}$

0 ft: $M = 0$

Figure 4-21*a* and *b* shows the moment diagram and the approximate shape of the elastic curve. The unloaded beam is horizontal at the wall; hence the elastic curve will be tangent to the original beam at the wall.

For convenience, Table 4-1 lists the values of the maximum moments for cantilever and simple beams when supporting the loadings indicated. Where more than one loading is applied to the beam, the total maximum moment can be obtained by adding the maximum moments for the separate loadings, provided the maximum moments occur at the same section.

TABLE 4-1

Type of beam	Loading	Maximum moment	Position of maximum moment
Cantilever	Uniform	$\dfrac{WL}{2}$	Wall
Cantilever	Conc. load at free end	PL	Wall
Cantilever	Conc. load at x from the free end*	$P(L - x)$	Wall
Simple	Uniform	$\dfrac{WL}{8}$	Center
Simple	Conc. load at the center	$\dfrac{PL}{4}$	Center
Simple	Two equal symmetrically placed conc. loads**	Px	Center (or at any section between the loads)

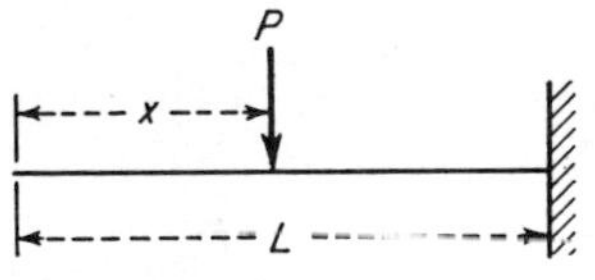

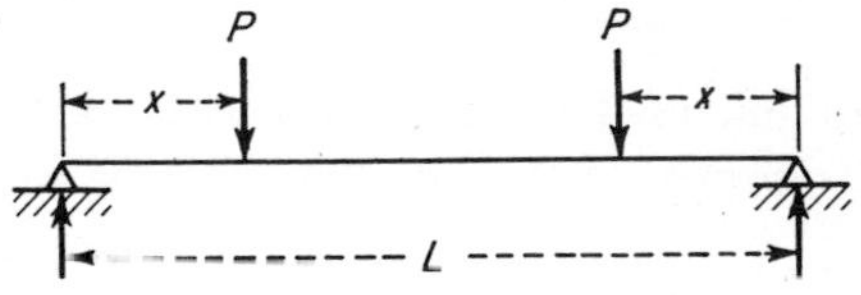

The symbol W is used to represent the *total* uniform load on the beam. The following example will illustrate this.

A beam is to support a uniform load of 120 lb per ft. If the beam is 16 ft long, what is the total uniform load?

$$W = 120 \times 16 = 1{,}920 \text{ lb}$$

Continuing this example, determine the maximum moment in the beam if the beam is supported at its ends. As this is a simple beam, from Table 4-1, the value of the maximum moment is given as $WL/8$. The desired units of the moment are inch-pounds, so the length L must be in inches. Then

$$M = \frac{WL}{8} = \frac{1{,}920 \times 192}{8} = 46{,}080 \text{ in.-lb}$$

If this beam also supports two symmetrically placed concentrated loads of 800 lb, placed 3 ft from each reaction, the maximum moment due to these loads will be

$$M = PX = 800 \times 36 = 28{,}800 \text{ in.-lb}$$

Since the maximum moments for both loadings occur at the same section, the moments calculated can be added. Thus the total maximum moment is

$$M = 46{,}080 + 28{,}800 = 74{,}880 \text{ in.-lb}$$

The use of Table 4-1 does not change the method of obtaining the values of the maximum moments as explained in this chapter but is intended to supplement the previously explained method.

4-8. Résumé. It is preferable to sketch the loading diagram, the shear diagram, the moment diagram, and the elastic curve directly below each other in the given order. The designer then has the complete picture of the values of V and M and can readily refer to them as well as to the elastic curve.

Figure 4-22*a* shows the loading diagram and the calculated reaction values. M is a couple with a value of 24 ft-kips counterclockwise, applied at the section 3 ft from the left reaction.

$$\Sigma M_{R_1} = 0 = +24 + 8R_2 - (12)8$$

$$R_2 = +9 \text{ kips (upward)}$$

$$\Sigma M_{R_2} = 0 = +8R_1 - 24 + (4)8$$

$$R_1 = -1 \text{ kip (downward)}$$

Figure 4-22*b* shows the plotted shear diagram.

0 ft_L: $V = 0$

0 ft_R: $V = -1$ kip

8 ft_L: $V = -1$ kip

8 ft_R: $V = -1 + 9 = +8$ kips

12 ft_L: $V = -1 + 9 = +8$ kips

12 ft_R: $V = -1 + 9 - 8 = 0$

Note that M does not cause a change in the shear diagram at the section where it is applied.

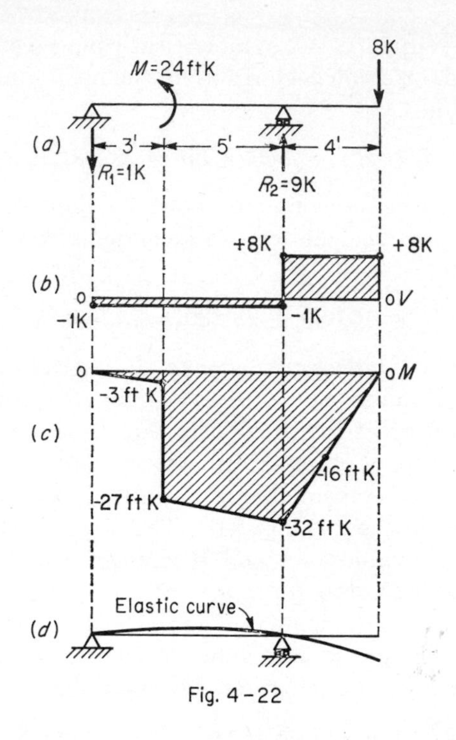

Fig. 4-22

Figure 4-22c shows the plotted moment diagram. In the process of obtaining the values of the moments, it is well to include moments at 3 ft_L and 3 ft_R. The moment M creates an abrupt change in the moment diagram.

0 ft: $M = 0$

3 ft_L: $M = -(3)1 = -3$ ft-kips

3 ft_R: $M = -(3)1 - 24 = -27$ ft-kips

8 ft: $M = -(8)1 - 24 = -32$ ft-kips

10 ft: $M = -(10)1 - 24 + (2)9 = -16$ ft-kips

12 ft: $M = -(12)1 - 24 + (4)9 = 0$

Figure 4-22*d* shows the elastic curve. Note that the moment diagram is completely negative and that the bending is entirely in one direction. The sections of zero shear were located without calculations; the sections of zero moment occurred at the ends of the beam; and there are no points of inflection.

Problems

In all problems:

(*a*) Sketch the loading diagram and solve the reactions and/or wall moment.

(*b*) Calculate the vertical shear values and plot the shear diagram. Locate the section(s) of zero shear.

(*c*) Calculate the moments and plot the moment diagram. Locate the section(s) of zero moment.

(*d*) Sketch the elastic curve.

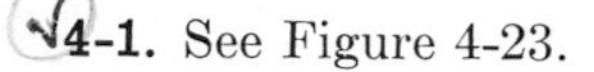

4-1. See Figure 4-23.

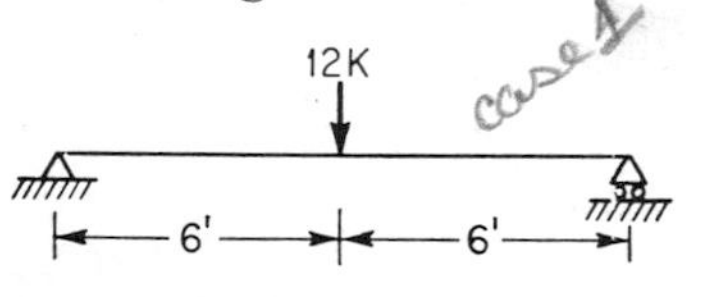

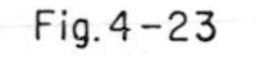

Fig. 4-23

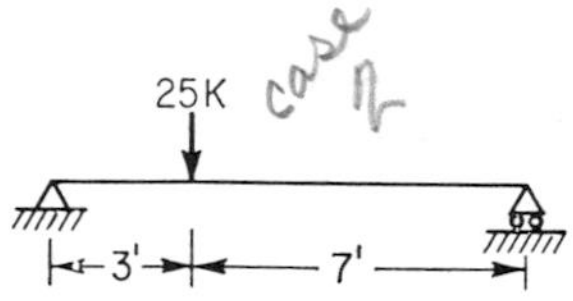

25K
3'
7'

Fig. 4-24

4-2. See Figure 4-24.

4-3. See Figure 4-25.

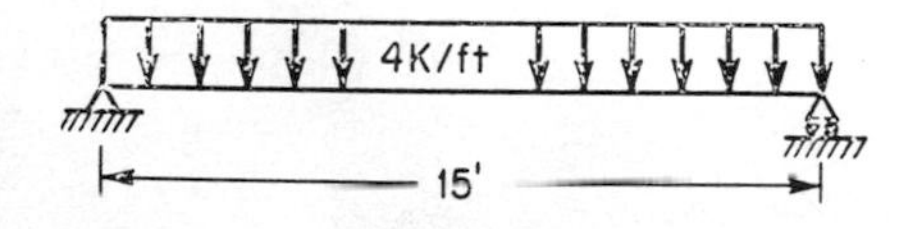

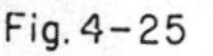

Fig. 4-25

1K/ft
6'
6'

Fig. 4-26

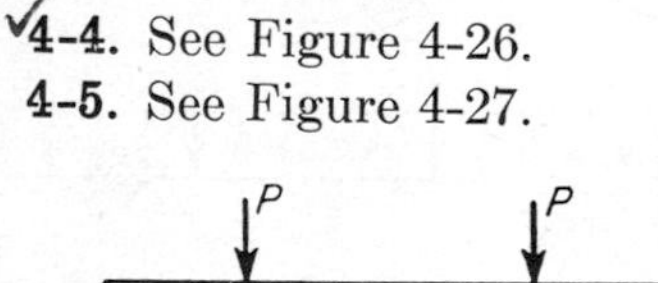

4-4. See Figure 4-26.

4-5. See Figure 4-27.

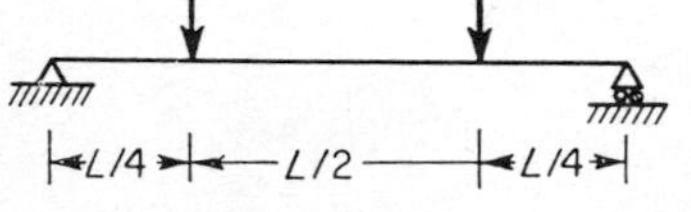

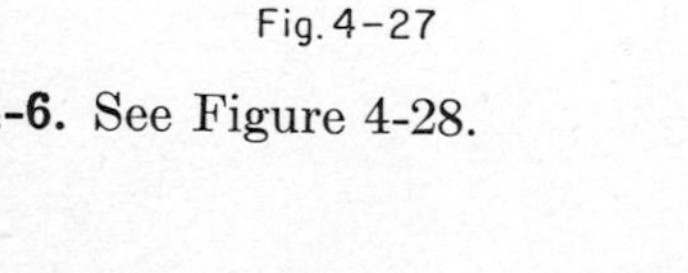

Fig. 4-27

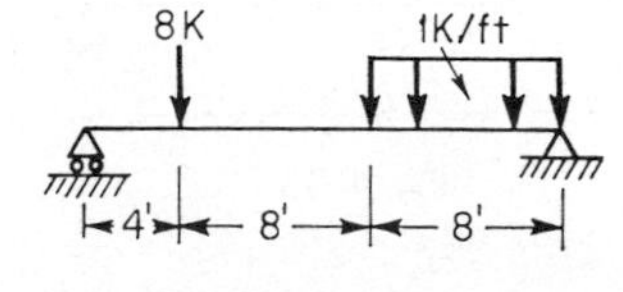

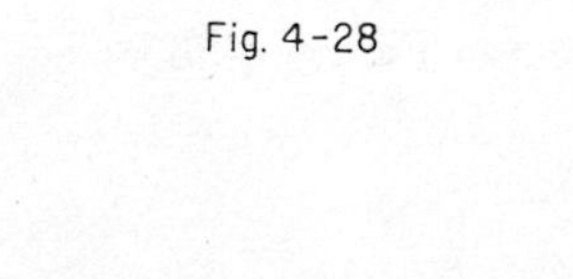

Fig. 4-28

4-6. See Figure 4-28.

4-7. See Figure 4-29.

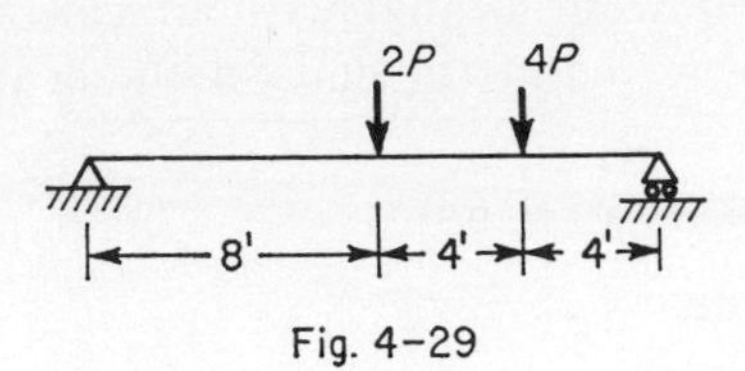

Fig. 4-29

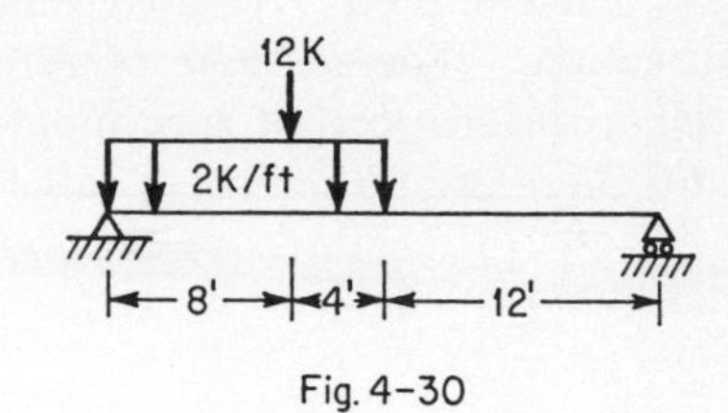

Fig. 4-30

4-8. See Figure 4-30.
4-9. See Figure 4-31.

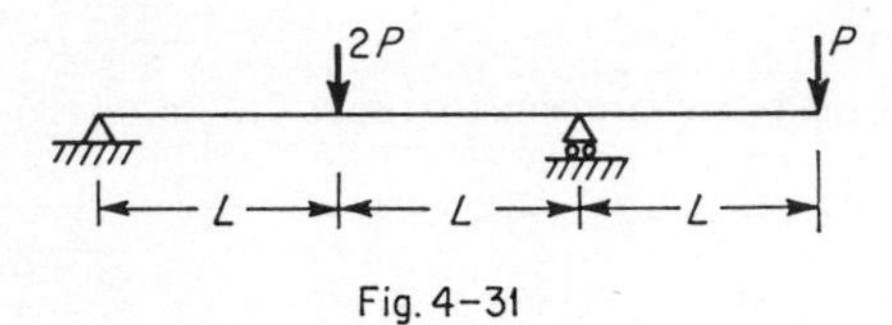

Fig. 4-31

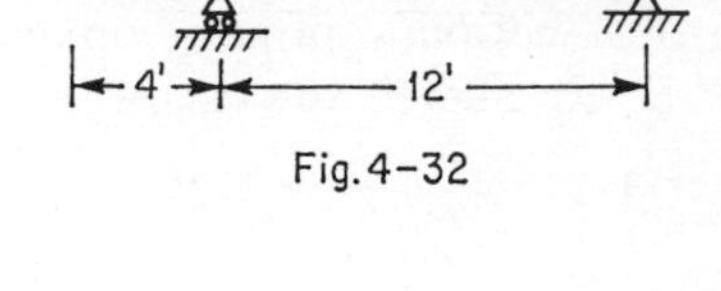

Fig. 4-32

4-10. See Figure 4-32.
4-11. See Figure 4-33.

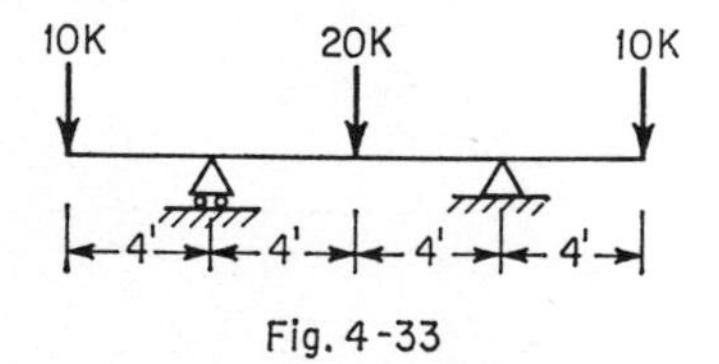

Fig. 4-33

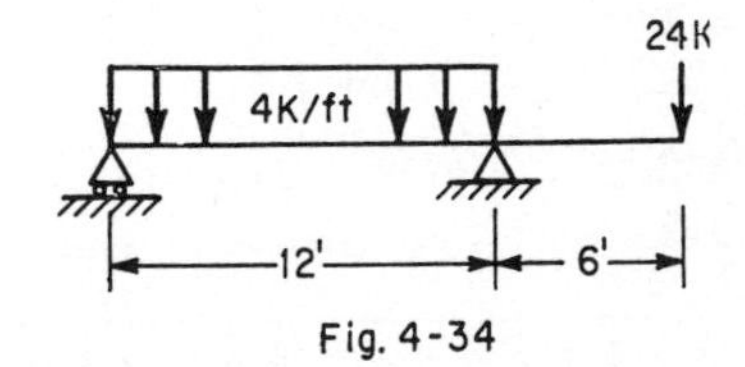

Fig. 4-34

4-12. See Figure 4-34.
4-13. See Figure 4-35.

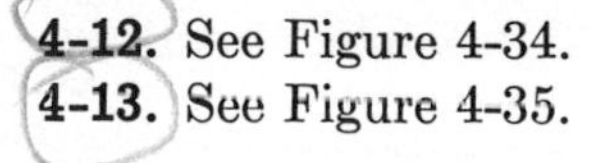
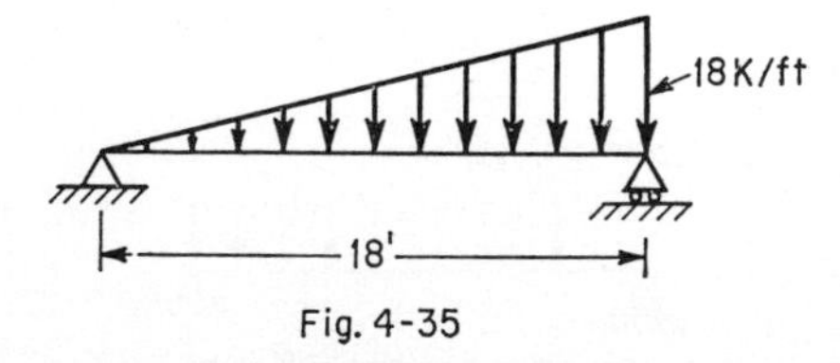

Fig. 4-35

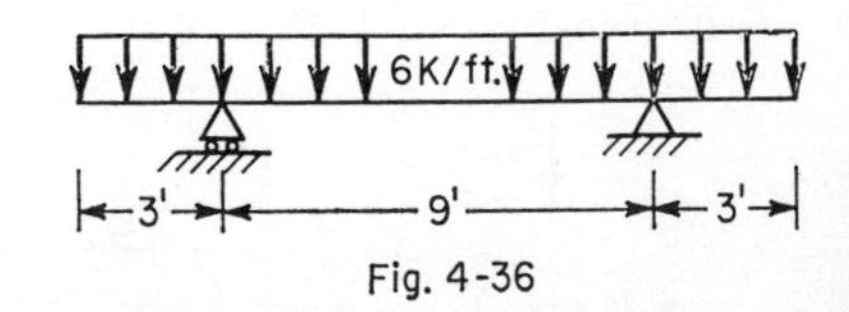

Fig. 4-36

4-14. See Figure 4-36.

4-15. See Figure 4-37.

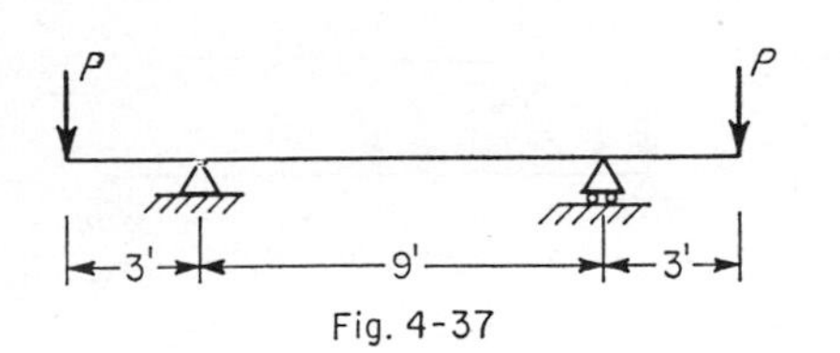

Fig. 4-37

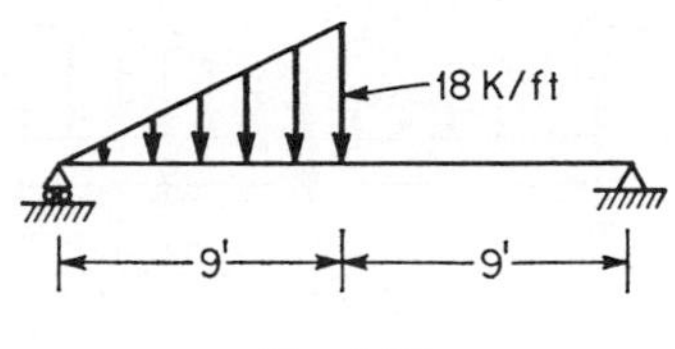

Fig. 4-38

4-16. See Figure 4-38.
4-17. See Figure 4-39.

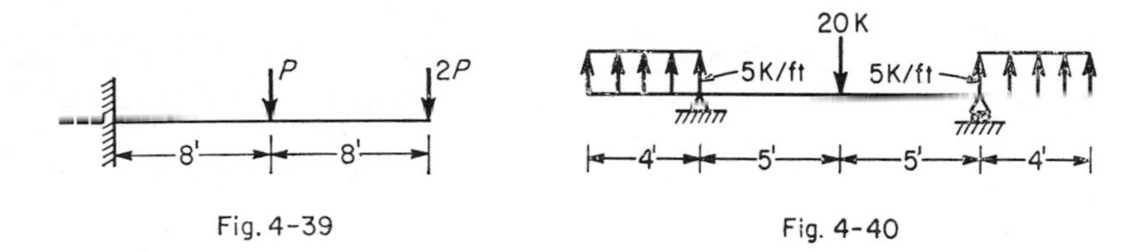

Fig. 4-39

Fig. 4-40

4-18. See Figure 4-40.
4-19. See Figure 4-41.

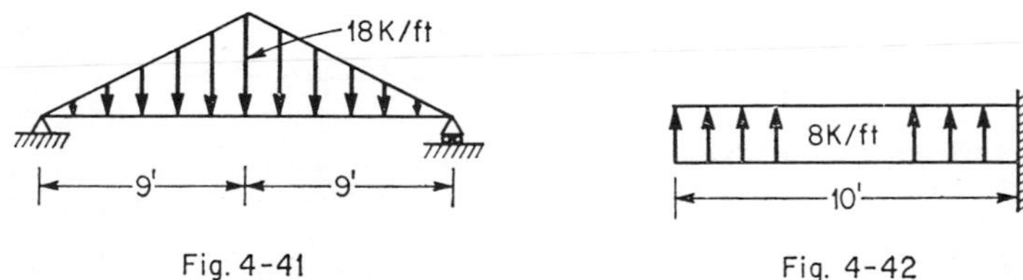

Fig. 4-41

Fig. 4-42

4-20. See Figure 4-42.
4-21. See Figure 4-43.

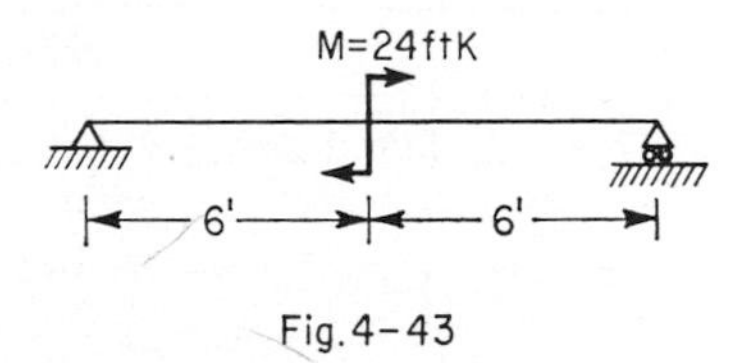

Fig. 4-43

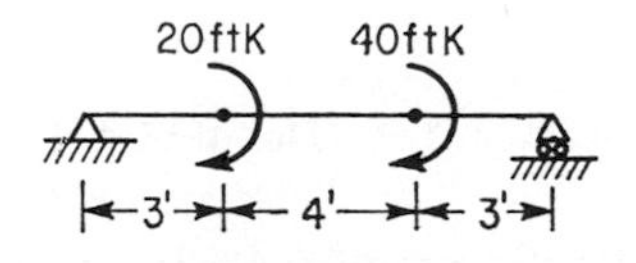

Fig. 4-44

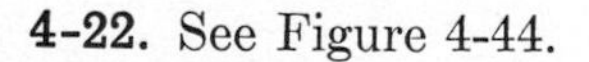
4-22. See Figure 4-44.

4-23. See Figure 4-45.

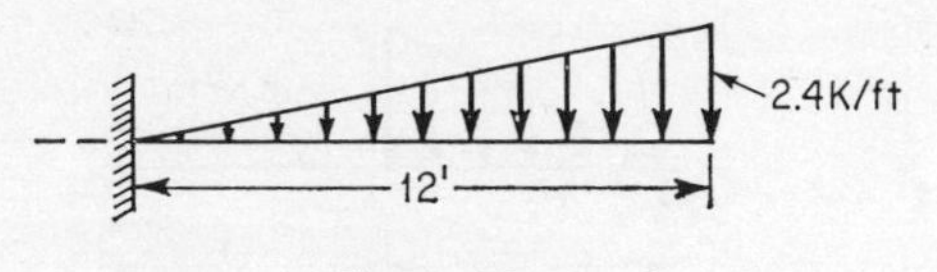

Fig. 4-45

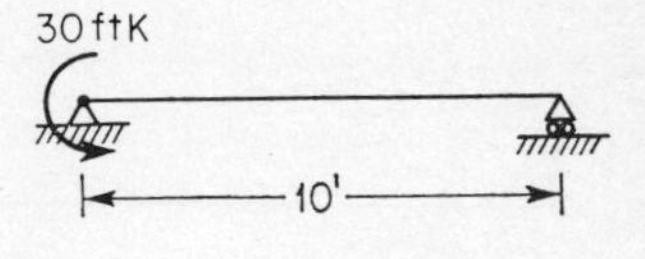

Fig. 4-46

4-24. See Figure 4-46.
4-25. See Figure 4-47.

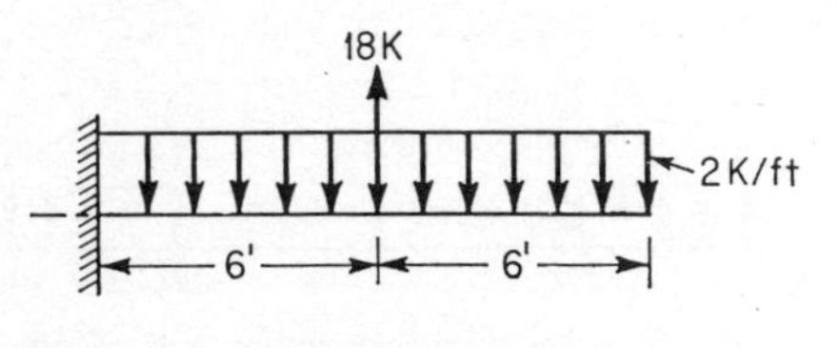

Fig. 4-47

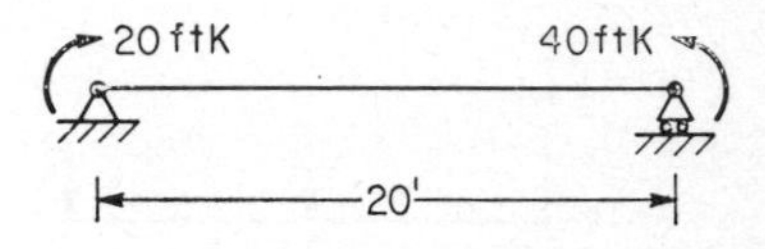

Fig 4-48

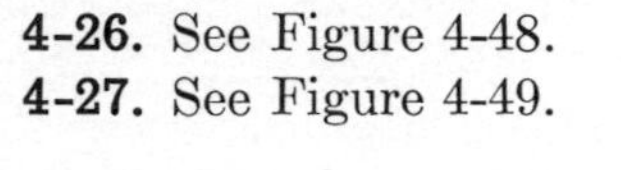

4-26. See Figure 4-48.
4-27. See Figure 4-49.

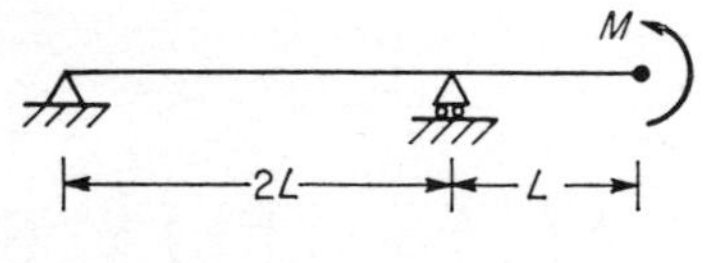

Fig. 4-49

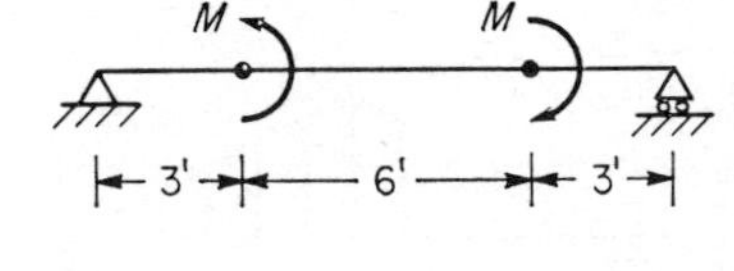

Fig. 4-50

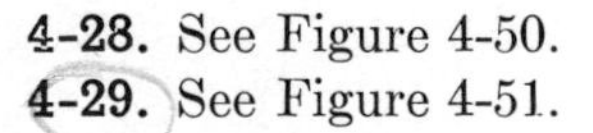

4-28. See Figure 4-50.
4-29. See Figure 4-51.

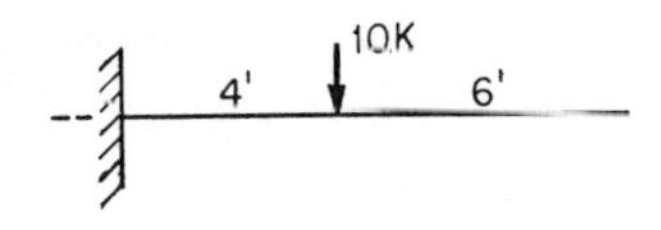

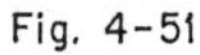

Fig. 4-51

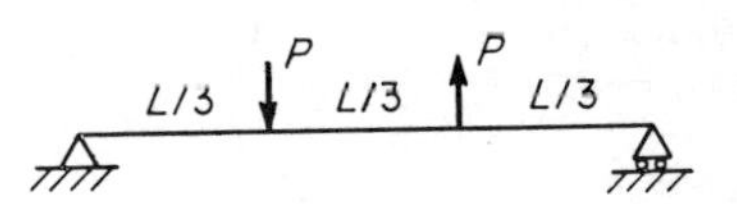

Fig. 4-52

4-30. See Figure 4-52.

CHAPTER 5
STRESSES IN BEAMS

5-1. Bending Stresses. In Chap. 4, it was shown that, when a beam was loaded in any manner, shear forces and bending moments were transmitted by each of the cross sections. It was also shown that a beam bends in the general direction(s) of the elastic curve. With the elastic curve as a basis, it should be observed that the beam material on the inside of the curve must shorten, and the material on the outside of the curve must stretch. Experimental methods show that the elastic curve does not change in length; therefore, the material at the curved plane through the beam coinciding with the elastic curve is not stressed. Experimental methods also show that the internal compressive and tensile stresses are proportional in magnitude to their respective distances from the curved plane of zero stress. This curved plane is known as the *neutral plane,* and its line on each beam cross section is known as the *neutral axis.* The neutral axis coincides with an axis which passes through the *centroid* of the cross section and which is parallel to the base of the cross section. Figure 5-1 shows a portion of a beam with the various curves and terms identified.

Certain basic assumptions are made which have been proved true by tests. They are:

1. A plane (cross) section remains plane.
2. The strain due to the bending is proportional to the distance from the neutral axis.
3. Stress is proportional to strain.

In Fig. 5-2*a* which shows only the cross section $ABCD$ of Fig. 5-1, the maximum compressive stress S_c occurs at the top of the beam, and the maximum tensile stress S_t will occur at the bottom of the beam. If we let S_c' be the compressive stress acting on a small particle of area dA at a distance of y from the neutral axis, then

$$\frac{S_c}{c_1} = \frac{S_c'}{y}$$

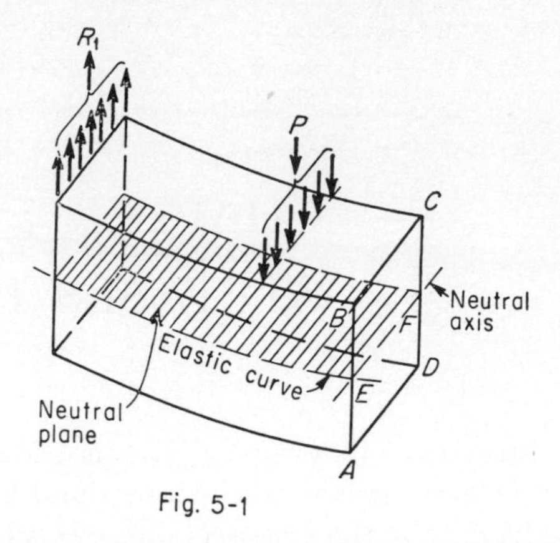

Fig. 5-1

Since we know that in general $P = SA$, the force P_c which acts on the particle of area equals $S_c' \, dA$. Also, for elastic conditions of stress, the external moment M must equal the internal moment

$$M = \Sigma P_c y = \Sigma S_c' y \, dA$$

Replacing S_c' by its equality $S_c y / c_1$,

$$M = \frac{\Sigma S_c y^2 \, dA}{c_1}$$

When the indicated summation process is completed by the use of advanced mathematics,

$$M = \frac{S_c I}{c_1} \qquad \text{or} \qquad M = \frac{S_c' I}{y}$$

where M = external moment (inch-pounds)
S_c = maximum compressive stress
S_c' = compressive stress at a distance y from the neutral axis
I = moment of inertia of the cross-sectional area about the neutral axis. *The neutral axis coincides with an axis which passes through the centroid of the cross-sectional area.* See the Appendix for values of I
c_1 = distance from the neutral axis E-F to the top of the beam (c_2 = distance from the neutral axis to the bottom of the beam)
y = distance from the neutral axis E-F to the point above or below the neutral axis where the stress (compressive or tensile) is to be calculated

The bending stress distribution is shown in Fig. 5-2*b*.

As an example of the application of the preceding theory, consider a simple beam 10 ft long supporting a single concentrated load of 4 kips at the midpoint. The beam is rectangular in cross section, 4 in. wide and 12 in. deep. Determine the maximum tensile and compressive stress in the beam.

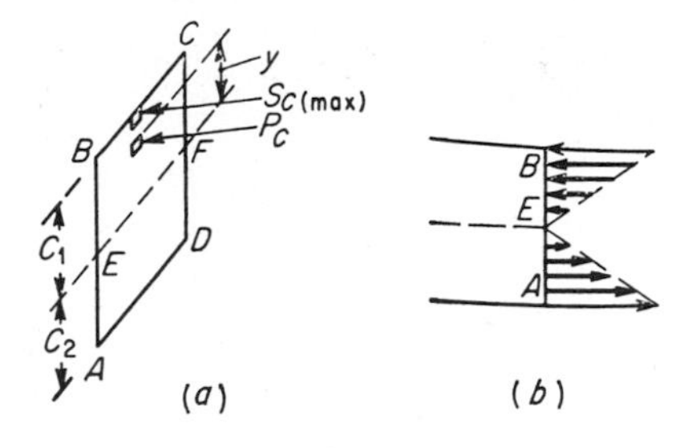

Fig. 5 2

From Table 4-1, the maximum moment

$$M = \frac{PL}{4}$$

and
$$M = +\frac{4{,}000(120)}{4} = +120{,}000 \text{ in.-lb}$$

From the Appendix,

$$I = \tfrac{1}{12}bh^3 = \tfrac{1}{12}4(12)^3 = 576 \text{ in.}^4$$

then
$$M = \frac{SI}{c} \qquad \text{and} \qquad S = \frac{Mc}{I}$$

$$S = \frac{120{,}000(6)}{576} = 1{,}250 \text{ psi} = S_t = S_c$$

Let us assume that the cross section of the beam in the preceding example is nonsymmetrical with respect to the neutral axis, and the corresponding values are: $I = 544$ in.^{4}; $c_1 = 3$ in.; $c_2 = 5$ in. Determining the maximum tensile and compressive stresses,

$$S_c = \frac{MC_1}{I} = \frac{120{,}000(3)}{544} = 662 \text{ psi}$$

and
$$S_t = \frac{MC_2}{I} = \frac{120{,}000(5)}{544} = 1{,}103 \text{ psi}$$

If the bending moment M had been negative, the magnitudes of S_c and S_t would be interchanged, with maximum S_c occurring at the bottom of the beam and maximum S_t at the top.

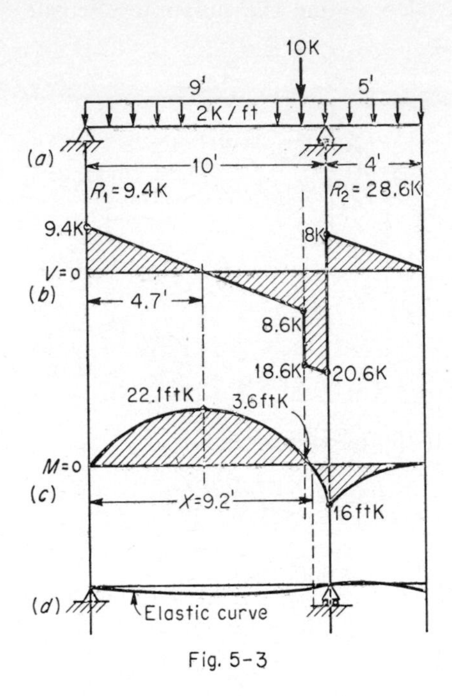

Fig. 5-3

An interesting example of a loaded beam is shown in Fig. 5-3*a*. Assuming that the reactions have been calculated and checked, the critical values of the vertical shear V are (Fig. 5-3*b*):

$$V_{O_R} = +9.4 \text{ kips}$$

$$V = 0 \qquad \text{at 4.7 ft from } R_1$$

$$V_{9_L} = -8.6 \text{ kips}$$

$$V_{9_R} = -18.6 \text{ kips}$$

$$V_{10_L} = -20.6 \text{ kips}$$

$$V_{10_R} = +8.0 \text{ kips}$$

$$V_{14} = 0$$

The critical values of the bending moments M are (Fig. 5-3c):

$$M_{V=0} = +9.4(4.7') - 2(4.7)\left(\frac{4.7}{2}\right) = +22.1 \text{ kip-ft}$$

$$M_9 = +9.4(9) - 2(9)(9/2) = +3.6 \text{ kip-ft}$$

$$M_{10} = M_4 \qquad \text{using the right end of the beam}$$

$$= -2(4)(4/2) = -16 \text{ kip-ft}$$

Since $M = 0$ between M_9 and M_{10}, the equation for locating the section of zero moment is:

$$+9.4x - 2(x)\frac{x}{2} - 10(x - 9) = 0$$

which reduces to

$$x^2 + 0.6x - 90 = 0$$

Solving for x,

$$x = +9.2 \text{ ft or } -9.8 \text{ ft}$$

An examination of the negative value for x indicates that it is imaginary since the location of the zero-moment section will be nearer to 9 ft than 10 ft from the left reaction.

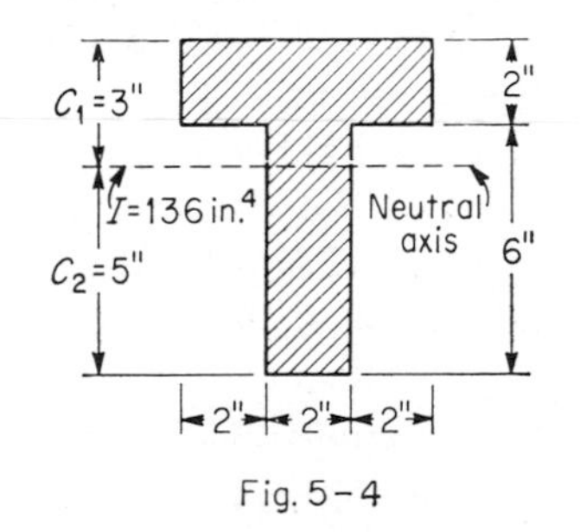

Fig. 5-4

The approximate elastic curve is sketched in Fig. 5-3d with a maximum positive moment of 22.1 ft-kips, and a maximum negative moment of 16 ft-kips. Using the beam cross section shown in Fig. 5-4, and calculating the maximum bending stresses for the positive bending,

$$S_c = \frac{22{,}100(12)3}{136} \qquad \text{and} \qquad S_t = \frac{22{,}100(12)5}{136}$$

from which

$$S_c = 5{,}850 \text{ psi} \qquad \text{and} \qquad S_t = 9{,}750 \text{ psi}$$

It is necessary to examine the maximum bending stresses for the negative bending:

$$S_t = \frac{16{,}000(12)3}{136} \qquad \text{and} \qquad S_c = \frac{16{,}000(12)5}{136}$$

from which

$$S_t = 4{,}240 \text{ psi} \qquad \text{and} \qquad S_c = 7{,}060 \text{ psi}$$

From the two maximum tensile stresses, it will be noted that the absolute maximum tensile stress occurs at the section of the beam which is 4.7 ft from R_1. Comparing the two maximum compressive stresses, the absolute maximum compressive stress of 7,060 psi occurs at the section of the beam 10 ft from the left end. The example indicates that maximum tensile and compressive stresses do not always occur at the same cross section of the beam. This can be true only when the cross section is nonsymmetrical with respect to the neutral axis. However, such a condition of two possible critical cross sections would also require a particularly careful analysis when the beam material has different allowable tensile and compressive strengths.

5-2. Design of Beams. In the preceding article, emphasis was placed on the calculation of the magnitudes and kind of bending stresses which were developed in the beam as the result of the application of an external bending moment.

In practice, generally the external bending moment and the allowable bending stress are known, and the problem of design is to select an economical rolled-steel shape. If the bending moment is small, the material selected for the beam may be timber or some other suitable structural material.

Rearranging the $M = SI/c$ equation,

$$\frac{I}{c} = \frac{M}{S}$$

In this form, it should be noted that I/c is a ratio dependent only on the dimensional properties of the cross section of the beam. This ratio is known as the *section modulus* of the beam.

The design and selection of structural steel beams are based on the specifications of the American Institute of Steel Construction. This institute is a nonprofit organization which exists for "the purpose of improving and advancing the use of fabricated structural steel through engineering, research and development, and promotion." These specifications have been widely adopted for municipal building codes, as well as for bridge codes.

In the Appendix of this text, a complete listing of the important design properties of all WF shapes is given. Continued reference will be made to these properties when the design requires the selection of a steel I or WF beam.

Since most I beams can be used in either of two general positions, as shown in Figs. 5-5 and 5-6, the tables give the section modulus S for each of these positions. Figure 5-7 shows the combined Figs. 5-5 and 5-6 as they appear at the top of each page in the Appendix. If the beam is used in the standing-up position (Fig. 5-5), the x-x axis is the

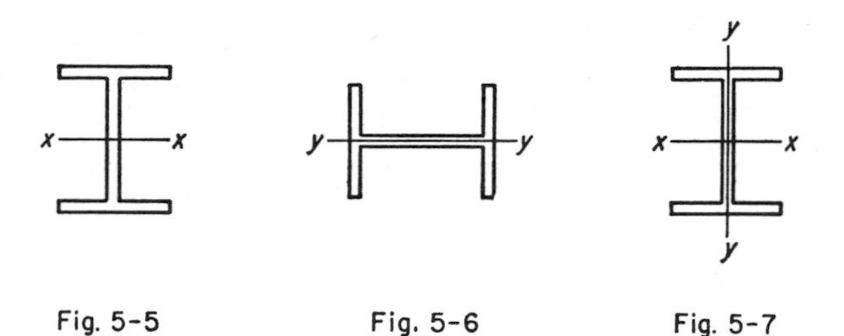

Fig. 5-5 Fig. 5-6 Fig. 5-7

neutral axis and the section modulus for the x-x axis will be used. If the beam is to be used with the neutral axis as the y-y axis, that section modulus will be used. Caution must be used against confusing the section-modulus symbol S which equals I/c and the stress symbol S.

The AISC suggests an allowable or design bending stress of 22,000 psi for common structural steel. In all problems and examples where a steel I beam is to be used to support a system of loads, the suggested stress of 22,000 psi will be used.

In selecting the most economical WF beam to support a certain loading, it is necessary to understand the listing of the available WF sizes and weights (per foot). A brief explanation of several of the columns of data follows:

Nominal size: this is the nominal depth in inches.

Weight per foot: note that within any group of beams with the same nominal depth, the weights per foot are listed in increasing magnitude from bottom to top.

Section modulus S for the axis x-x: these are listed in increasing magnitude from bottom to top in any group of beams.

An example of the selection of an I beam will illustrate the proper procedure. It should be kept in mind that steel is purchased by the pound. Hence, unless there is a limiting specification other than the stress, the minimum-weight beam with a section modulus equal to or greater than that required is the correct selection.

A simple beam 15 ft long is to support a uniform load of 20 kips per ft. What size I beam is required?
From the table of maximum moments

$$M = \frac{WL}{8} = \frac{300{,}000(\text{lb}) \times 180(\text{in.})}{8} = 6{,}750{,}000 \text{ in.-lb}$$

This should be verified by solving the reactions and by calculating the V and M diagrams. Then

$$S = \frac{I}{c} = \frac{M}{S} = \frac{6{,}750{,}000}{22{,}000} = 307 \text{ in.}^3$$

Checking the column of data giving the axis x-x section moduli, proceeding from the bottom of the list upward and *excluding* the nominal 14 × 14½ and 14 × 16 sizes which are column sizes, the safe beams are:

Size	Wt/ft	S
21 × 13	142	317.2
24 × 14	130	330.7
30 × 10½	116	327.9

Only the lightest *safe* beam in any size group is listed. Those with values of I/c obviously too large are not considered because of the excess weight. Of the three beams listed above, the correct selection is the 30 × 10½ beam, and it is specified: "Use 30 WF at 116 lb." The algebraic sign of the bending moment M is disregarded in the selection of the WF beams. The weight of the beam is included in the uniform loading or is assumed to be negligible for other types of loading.

The safe-design stress for timber beams is 1,000 psi, and the timber is usually available in rectangular shapes. Rough-sawed timber is full size, and dressed or surfaced timber is ⅜ in. smaller in each dimension than rough-sawed timber. A rough-sawed 4 in. × 12 in. is 3⅝ in. × 11⅝ in. dressed. In order to simplify the calculations, all timber beams will be assumed to be full size. Moments of inertias and section moduli of common shapes are listed in the Appendix.

A simple timber beam which is 12 ft long with a depth of 12 in. is to be used to support two symmetrically placed, equal, concentrated loads. Each load P is 3,000 lb, and a load is placed 3 ft from each end of the beam. Determine the required width b of the beam.

The following calculated values should be carefully checked and verified:

$R_1 = R_2 = 3{,}000$ lb

Maximum $V = \pm 3{,}000$ lb

$V = 0$ from 3 ft to 7 ft from R_1

Maximum $M = +108{,}000$ in.-lb at all sections from 3 ft to 7 ft from R_1

Letting $S = 1{,}000$ psi, then

$$108{,}000 = \frac{1{,}000(b)12^3}{(6)12}$$

from which

$$b = 4.5 \text{ in.}$$

5-3. Shearing Stress in Beams. If we were to use a number of 2-in. × 6-in. timber planks as a simple beam (Fig. 5-8), considering

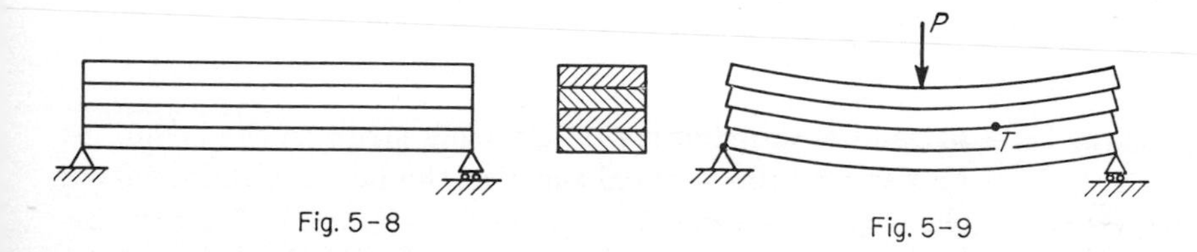

Fig. 5-8

Fig. 5-9

them weightless, and then apply a vertical load P of sufficient magnitude to cause considerable sag in the beam shown in Fig. 5-9, we would see that the adjacent surfaces of any two of the planks would slide or shear with respect to each other. This is particularly true of a solid-timber beam where the material has natural planes of cleavage.

Considering a minute particle T of the beam, and enlarging it as shown in Fig. 5-10, we see that the shearing stresses developed on the horizontal surfaces $EFBA$ and $HGCD$ of the particle can act in the directions shown. If these stresses were each multiplied by the equal areas on which each acts $[P_s = A_{EFBA}(S_s) = A_{HGCD}(S_s)]$, the resulting forces would cause the particle to rotate clockwise. Since the particle does not rotate, equal and opposite forces P_s and P_s are developed on the left and right faces of the particle which provide the counterclockwise moment to prevent rotation of the particle. Dividing the vertical forces by their equal areas produces equal and opposite shearing stresses S_s on the side faces. Thus, the horizontal and vertical shearing stresses at any point in the beam material are equal.

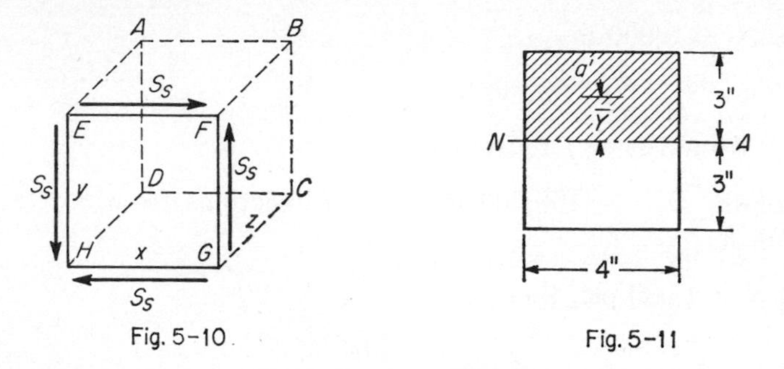

Fig. 5-10

Fig. 5-11

5-4. Horizontal Shearing Stress. The formula by which the maximum unit horizontal shearing stress (and vertical shearing stress) is calculated is derived by a process involving considerable advanced mathematics, so it will suffice to give the formula and define the terms involved.

$$S_s = \frac{V}{It} a'\bar{y}$$

where S_s = maximum unit horizontal shearing stress in the beam, psi
V = maximum total vertical shear in the beam, obtained from the shear diagram, lb
I = moment of inertia of the cross section of the beam about the neutral axis (centroid)
t = width of the beam at the *neutral axis*
a' = area of the beam above or below the neutral axis
$\bar{y}$ = distance from the *neutral axis* to the *centroid* of the a' area

The preceding formula for determining the maximum horizontal shearing stress can also be used to calculate the shearing stress at any point on any vertical cross section of the beam. By using the value of the total vertical shear V at a particular section, the maximum shearing stress for that section is obtainable.

If the shearing stress is desired at a particular point above or below the neutral axis, a' will be the area from that point at which the stress is desired to the *top* or *bottom* of the beam; $\bar{y}$ will then be the distance from the neutral axis to the centroid of the a' area.

The shearing stress distribution on a cross section of the beam is parabolic, with the *maximum shearing stress at the neutral axis and zero stress at the top and bottom of the beam.*

Example. A timber beam 4 in. × 6 in. 10 ft long is used to support a concentrated load of 3,000 lb at the center of the beam. The beam is supported at the ends. What is the maximum unit horizontal shearing stress in the beam? See Fig. 5-11.

By plotting a shear diagram, it will be seen that the maximum value of the total vertical shear V is 1,500 lb. Then

$V = 1{,}500$ lb

$I = \frac{1}{12}bh^3 = \frac{1}{12}4(6)^3 = 72$

$t = 4$ in.

$a' =$ area above the neutral axis

$= 4 \times 3 = 12$ sq in.

$\bar{y} =$ distance from the neutral axis to the centroid of the area $= \frac{1}{2}(3) = 1\frac{1}{2}$ in.

$$S_s = \frac{V}{It} a'\bar{y}$$

$$= \frac{1{,}500}{72 \times 4}(12)(1\tfrac{1}{2})$$

$$= {}^{1500}\!/_{16} = 93.75 \text{ psi}$$

(It follows that the maximum unit vertical shear is 93.75 psi. That is, the beam has equal horizontal and vertical shear stresses. However, the timber is easier to shear along the grain, which is normally placed horizontally.)

It is often customary in beams of *rectangular cross sections* to calculate the maximum unit horizontal shear by reducing the full equation to

$$S_s = \frac{V}{It} a'\bar{y}$$

$$= \frac{V}{(bh^3/12)b}\left(b\,\frac{h}{2}\right)\left(\frac{h}{4}\right)$$

$$= \frac{3}{2}\,\frac{V}{bh}$$

The term V/bh represents the *average* unit vertical shear in the beam; hence, $\frac{3}{2}$ of the average unit vertical shear in a beam of *rectangular cross section* equals the *maximum unit horizontal shearing stress.*

Applying this to the preceding example,

$$V = 1{,}500 \text{ lb}$$

$$bh = 4 \times 6 = 24 \text{ sq in.}$$

$$S_s = \frac{3}{2}\frac{V}{bh} = \frac{3 \times 1{,}500}{2 \times 24} = \frac{1{,}500}{16} = 93.75 \text{ psi}$$

5-5. Shearing Stresses in I Beams. The usual assumption made in the case of steel I beams, channels, and T beams is that the total vertical shear is carried entirely by the web of the beam, the web extending the *full height* of the beam. While the flanges offer some little resistance to vertical shear, the preceding assumption is on the side of safety. Thus, the *average* vertical web shearing stress

$$\text{Average } S_s = \frac{V}{\text{web area}} = \frac{V}{td}$$

where t is the thickness of the web and d is the nominal *full* height of the beam.

Example. A cantilever beam 6 ft long is to support a uniform load of 4,000 lb per ft. What size steel I beam is required? Is the web safe if the allowable average vertical shear is 14,500 psi, as specified by the American Institute of Steel Construction?

By drawing the shear and moment diagrams, maximum $V = 24{,}000$ lb, and the maximum bending moment $M = 72{,}000$ ft-lb.
Using an allowable bending stress of 22,000 psi,

$$M = S\frac{I}{c}$$

$$72{,}000 \times 12 = 22{,}000\frac{I}{c}$$

$$\frac{I}{c} = 39.3 \text{ in.}^3 \text{ required section modulus}$$

From the table of WF beams, the lightest I beam with a section modulus greater than 39.3 is a 12 WF at 31 lb. The web thickness of this I beam is 0.265 in., and the depth is 12 in. The total web area is

0.265 × 12 in. Then

$$\text{Average } S_s = \frac{V}{td} = \frac{24{,}000}{0.265 \times 12} = 7{,}550 \text{ psi}$$

As this is less than the safe allowable average unit vertical shear for steel of 14,500 psi, the beam is safe.

Problems

5-1. A timber beam with a height of 12 in. is used to support the loading of Prob. 4-1. If the allowable bending stress is 1,500 psi, what width beam is required?

5-2. A WF beam which has a nominal depth of 10 in. is used as the beam for Prob. 4-2. Determine: (*a*) the minimum weight per foot of the beam; (*b*) the maximum unit web shearing stress.

5-3. The WF beam used to support the loading on the beam of Prob. 4-3 has a maximum limiting nominal depth of 18 in. What is the recommended WF section?

5-4. A timber beam with the width equal to one-half of the depth is used for the beam of Prob. 4-4. What are the required actual dimensions of the beam cross section if the allowable bending stress is 1,000 psi?

5-5. What size WF beam is required for the loading given in Prob. 4-5? Use P = 20 kips and L = 16 ft.

5-6. A structural steel T beam has the following design properties: A = 12.4 in.^{2}; I_x = 166 in.^{4}; c_1 = 2.97 in.; c_2 = 9.08 in.; wt per ft = 42 lb. Is this beam safe to support the loading given for Prob. 4-6?

5-7. A structural steel shape has the following design properties: A = 14.8 in.^{2}; I_x = 445 in.^{4}; c_1 = 4.56 in.; c_2 = 9.58 in.^{2}; wt per ft = 50.7 lb. In Prob. 4-9, use P = 16 kips and L = 5 ft. What are the maximum tensile and compressive stresses developed in the beam?

5-8. Referring to Prob. 4-10, what minimum-weight WF steel beam would you recommend if the beam is to be placed with the web in a *horizontal* position? The minimum nominal depth (the horizontal dimension as the beam is used) shall not be less than 12 in.

5-9. Determine the maximum tensile stress in a steel beam to be used for the loading in Prob. 4-11. The beam has the following design properties: A = 19.1 in.^{2}; I_x = 223 in.^{4}; c_1 = 2.47 in.; c_2 = 9.66 in.; wt per ft = 65 lb.

*5-10. Using the beam loadings of Prob. 4-12, determine the minimum-weight 18-in. WF beam which can be safely used. Is this beam safe if the design limit is 14,500 psi maximum web shearing stress?

5-11. Will a 24-in. WF beam at 94 lb be safe to support the nonuniform loading on the beam of Prob. 4-13?

5-12. An 18-in. WF-at-70-lb beam is to be used to support the loading on the 15-ft-long beam of Prob. 4-15. What is the maximum force P which can be applied?

5-13. An aluminum-alloy I beam has the same dimensional design properties as a 12-in. WF-at-50-lb beam. If the loading on the beam of Prob. 4-18 is used, what maximum bending stress is developed in the aluminum?

5-14. A timber beam is 4 in. wide and 12 ft long. It is used as a simple beam to support a uniform load of 200 lb per lineal ft and a concentrated load of 1,000 lb at the midpoint of the span length. With a safe bending stress of 1,000 psi, what *actual* height of beam is required? What maximum shearing stress (horizontal) is developed in this actual-size beam?

*5-15. A simple beam 3 ft long supports a uniform load of 100,000 lb per ft. What is the maximum average shearing stress in the web of a 21-WF-at-62-lb beam? Would this beam be considered safe according to a maximum allowable web shearing stress of 14,500 psi?

5-16. A 4-in.-wide × 12-in. timber beam is used to support the loading in Prob. 4-17. If the allowable bending stress for the timber is 1,000 psi, what is the maximum value of P? What maximum shearing stress is developed in this beam?

5-17. A rectangular timber beam, with a maximum shearing stress of 50 psi, is used to support the loading of Prob. 4-21. The depth of the beam is specified to be 12 in. What is the maximum bending stress in the beam?

5-18. A steel beam has the following design properties: $A = 33.7$ in.^{2}; $I_x = 4{,}110$ in.^{4}; $c_1 = 11.4$ in.; $c_2 = 15.8$ in.; wt per ft $= 115$ lb. What maximum compressive stress is developed in this beam if it is loaded as shown in Prob. 4-20?

5-19. A T beam has the following dimensional properties (see Fig. 5-12): $c_1 = 4$ in.; $c_2 = 2$ in.; $I_x = 64$ in.^{4} Using the loading from Prob. 4-22, determine the maximum shearing stress at a point just *above* the neutral axis. What is the maximum shearing stress just *below* the neutral axis?

5-20. Using all the pertinent information from Prob. 5-19, determine the maximum tensile bending stress in the T beam. What is the

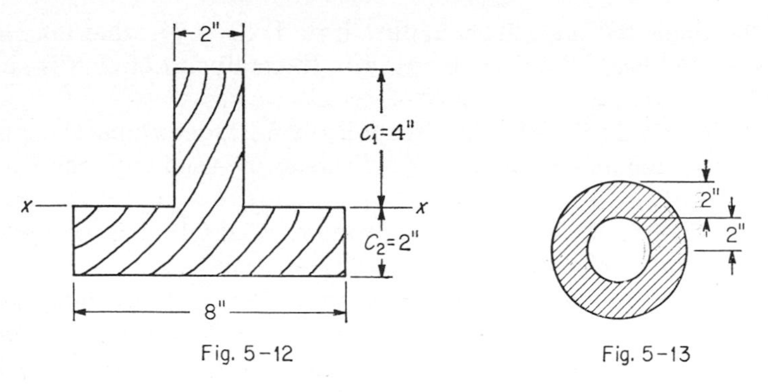

Fig. 5-12

Fig. 5-13

maximum compressive stress in the beam? What is the maximum bending stress at a section of the beam which is 2.99 ft from the left reaction?

5-21. The tubular cross-sectional beam shown in Fig. 5-13 is used to support the loads given in Prob. 4-24. Determine the maximum bending stress at the outside surface and the corresponding bending stress at the inside surface of the beam.

5-22. A beam has a symmetrical hollow rectangular cross section as shown in Fig. 5-14. Using the beam load for Prob. 4-23, determine the maximum bending stress and the bending stress at point A.

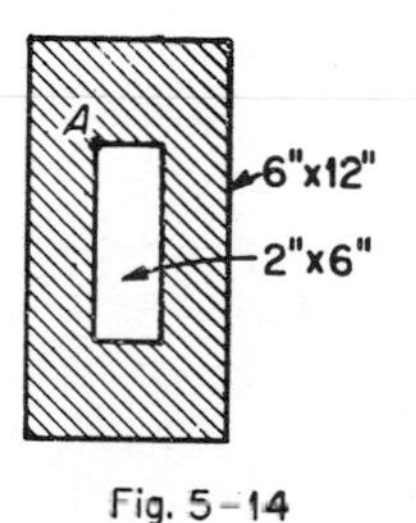

Fig. 5-14

5-23. Using the beam cross section and other necessary data from Prob. 5-22, determine the maximum shearing stress in the beam.

5-24. Two 6-in. × 6-in. × 1-in. steel angles are rigidly attached to each other to form a T beam with the 2-in.-thick stem (or web) in the vertical position and the flange 12 in. wide at the top. Obtain the design properties from the table of Properties of Equal Angles in the Appendix.

If the T beam is used as a simple beam 8 ft long to support a uniform load, determine the maximum permissible uniform load in pounds per foot.

5-25. Using the same T-beam data from Prob. 5-24, what maximum concentrated load P could you apply to the midpoint of an 8-ft simple beam?

5-26. Using the T-beam data from Prob. 5-24, determine the magnitude of the maximum moment M which can be used with the loading of Prob. 4-28.

5-27. Determine the required minimum-weight 10-in. or deeper WF steel beam to safely support the loading of Prob. 4-29.

5-28. An aluminum-alloy beam has the following dimensional design properties: $A = 7.36$ in.^{2}; $I_x = 53.9$ in.^{4}; $c_1 = 2.14$ in.; $c_2 = 6.86$ in. The maximum allowable bending stress is 10,000 psi. For Prob. 4-30, determine the maximum value of P when $L = 12$ ft.

5-29. If a 21-in. WF-at-68-lb steel beam is used to support the loading system given in Prob. 4-16, determine the maximum average web shearing stress.

5-30. A steel beam has the following design properties: $A = 28.4$ in.^{2}; $I_x = 2{,}730$ in.^{4}; $c_1 = 9.8$ in.; $c_2 = 14.4$ in.; depth $= 24.2$ in.; web $t = 0.44$ in. The maximum allowable bending stress is 22,000 psi, and the maximum web shearing stress is 14,500 psi. If the structural shape is used as a simple beam supporting a concentrated load P at the midpoint, determine the maximum load P and the maximum usable length L.

CHAPTER 6
BEAM DEFLECTIONS

6-1. Bending. One of the many important things that was discussed in Chap. 4 concerned the direction of bending of the neutral axis. This curve, which is called the *elastic curve*, assumes a definite direction of curvature with respect to the original straight-line position of the beam due to the bending moments at the many cross sections of the beam. The relative position of the curve perpendicular to the original straight-line position of the beam at any point along the curve is called the *deflection* of the beam. There is some small magnitude of deflection due to the shearing stress in the beam, but being very small compared to the deflection due to the bending moment, it will be neglected. This should not be construed to imply that beam deflections are large in magnitude.

An elastic curve is a smooth, continuous curve without any abrupt changes in the directions of curvature, and the *radius of curvature* at a specific point on the elastic curve is the physical measurement of the geometrical radius at that point.

The *slope* of the elastic curve at any point is the trigonometric slope of, or the tangent of the angle of, a straight-line tangent to the elastic curve at that point. The angle of slope is measured with respect to the horizontal original straight-line *unloaded* position of the elastic curve of the beam. Each beam is assumed to be placed in the horizontal position, and the elastic curve has zero deflection at the support(s).

6-2. Radius of Curvature. In Fig. 6-1, the side view of a portion of a beam with negative bending is shown. The bending is highly exaggerated, and the short length dx of the beam is measured along the curve of the neutral axis or elastic curve. For practical consideration, this length can be considered straight, and all linear deformations (stretch at the top of beam and shortening at the bottom of the beam) are assumed to accumulate at the right side of the dx length of the beam being considered.

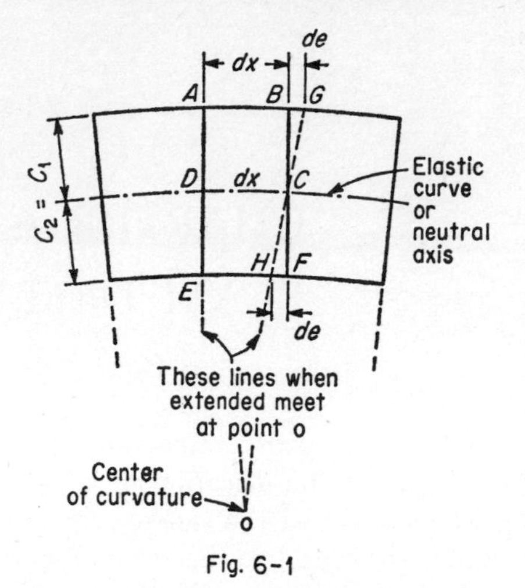

Fig. 6-1

Before the beam is loaded, $dx = AB = DC = EF$ and the beam is straight. After loading, AB is increased by de so that

$$AG = AB + BG = dx + de$$

and with $c_1 = c_2$

$$EH = EF - HF = dx - de$$

From the geometry of the figure

$$DC:DO::BG:BC::HF:FC$$

Then

$$\frac{DC}{DO} = \frac{BG}{BC} \quad \text{and} \quad \frac{dx}{\rho} = \frac{de}{c_1}$$

Similarly

$$\frac{DC}{DO} = \frac{HF}{FC} \quad \text{and} \quad \frac{dx}{\rho} = \frac{de}{c_2}$$

where ρ (the Greek letter rho) is the radius of curvature of the beam at the neutral axis or elastic curve when $c_1 = c_2$. The preceding relationships are likewise true for conditions when c_1 is not equal to c_2.

Recalling that for tensile or compressive axial stress,

$$\varepsilon = \frac{e}{L}$$

and from the above

$$\frac{de}{dx} = \varepsilon = \frac{c}{\rho}$$

Also, for axial tensile or compressive stress

$$E = \frac{S}{\varepsilon}$$

and for bending (tensile or compressive) stress

$$S = \frac{Mc}{I}$$

then
$$\varepsilon = \frac{c}{\rho} = \frac{S}{E} = \frac{Mc}{IE}$$

and canceling the c terms in the second and fourth parts of the equalities,

$$\rho = \frac{EI}{M}$$

All the above development assumes, and is valid only for, elastic stress conditions.

An example of the stress due to bending is that of a perfect ring being formed from a strip of steel 1.2 in. wide and 0.01 in. thick. The ring has a mean radius of 6 in. and the depth dimension of the ring is its thickness. From

$$M = \frac{EI}{\rho} = \frac{30 \times 10^6 \times 1.2(0.01)^3}{6(12)}$$

$$= 0.5 \text{ in.-lb}$$

$$S = \frac{Mc}{I} = \frac{0.5(0.01)(12)}{2(1.2)(0.01)^3} = 25{,}000 \text{ psi}$$

which would be less than the elastic limit of the steel.

6-3. Beam Deflections. Since the deflection of a beam has already been defined in Art. 6-1, it becomes necessary only to interpret graphically the physical concepts of the elastic curve. Figure 6-2 shows the highly exaggerated bending of a *symmetrically* loaded simple beam. For symmetrical loading, the elastic curve bends symmetrically with respect to the midpoint of the span length L, and the maximum deflection Δ (the Greek letter delta) occurs at the midpoint of the span.

The deflection at any other point on the elastic curve is designated by y. The *slope* of the elastic curve at any point is the tangent of the angle Θ (the Greek letter theta) which a line tangent to the elastic curve, at that point, makes with a line tangent to the elastic curve at its lowest (or horizontal) point. This definition is valid only for beams which are initially straight and are in a *horizontal* position. (This is consistent with the previous statements concerning the placements of beams.)

For a cantilever beam, the elastic curve has zero slope at the fixed (wall) end, with a maximum deflection at the free end.

The principle of superposition is particularly useful in determining the deflection of a beam subjected to more than one single load or type

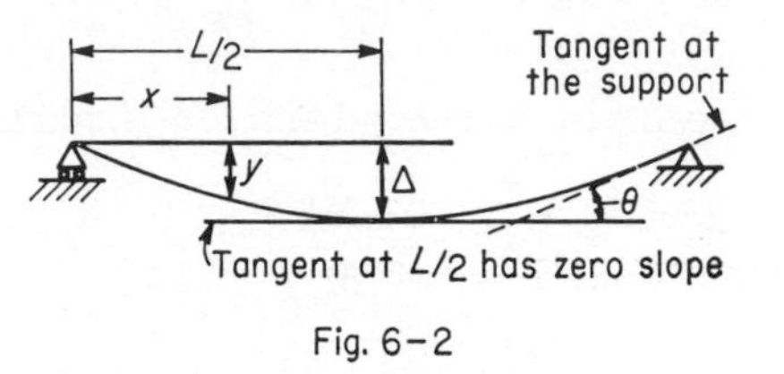

Fig. 6-2

of loading. Thus, the maximum deflection of a simple beam subjected to a concentrated load at the midpoint *and* to two other symmetrically placed, equal, concentrated loads is equal to the sum of the maximum deflection due to the midpoint loading *and* the maximum deflection due to the pair of symmetrically placed equal loads. This is valid only when the Δ terms occur at the same point on the elastic curve.

Figure 6-3 shows an application of the principle of superposition as it can be used for the preceding example. If Δ_1 equals the maximum deflection due to the midpoint load, and Δ_2 equals the maximum deflection due to the pair of symmetrically placed equal loads ($P_1 = P_3$), then

$$\Delta_{\text{max}} = \Delta_1 + \Delta_2$$

If P_1 and P_3 pull upward, algebraically

$$\Delta_{\text{max}} = \Delta_1 - \Delta_2$$

For the few possible cases where there is a portion of the length of the beam for which the moment M diagram is *zero,* that portion or length of the elastic curve is *straight.* The proof of this comes from the equation for the radius of curvature ρ of the elastic curve

$$\rho = \frac{EI}{M}$$

When the bending moment M is zero, ρ equals infinity and the successive radii are parallel; hence the beam is straight.

There are several methods by which it is possible to develop expressions for the deflection and slope at any point on an elastic curve. However, all these methods require an understanding of mathematics beyond the scope and level of that required for an understanding of the theory which is developed in this text. Thus, it is impossible to derive these values of y, Δ, and the $\tan \Theta$. For this reason, Table 6-1, which contains sufficient data for the determination of the slopes and

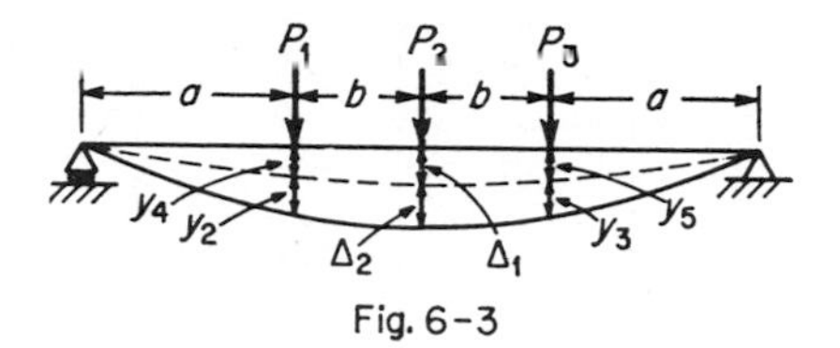

Fig. 6-3

deflections, will suffice to make available the numerical magnitudes for many of the simpler cases of beam loadings. It will be necessary to refer to the table, but the use of moment diagrams and their accompanying elastic curves will materially aid in a better understanding of the significance of the calculated results.

Deflections of beams are, numerically, very small. This is illustrated by the time-honored maximum allowable deflection of a simple beam which is limited to $\frac{1}{360}$ of the span or length of the beam. For a simple beam of length = 12 ft = 144 in., the allowable

$$\Delta = \tfrac{1}{360}(144) = 0.40 \text{ in.}$$

It follows that if the deflection is small, the slope of the elastic curve at any point is small, and the $\tan \Theta$ is equal to the angle Θ or $\tan \Theta = \Theta$. This can be checked by noting that for small angles, the

$$\sin \Theta = \tan \Theta = \Theta$$

When any of the slope equations listed in Table 6-1 are used, it is advisable to replace the magnitudes of x and/or L in *foot* units, and then multiply the results by 144 in order to ensure the correct units of Θ.

Table 6-1. Slopes and Deflections of Elastic Curves for Statically Loaded Beams

Loading	Θ = slope	Position of Θ	y	Position of y	Δ max	Position of max
P; L/2; L/2; X	$\frac{PL^2}{16EI}$	At end	$\frac{Px}{48EI}(3L^2 - 4x^2)$	At x from reaction	$\frac{PL^3}{48EI}$	At load
P; P; X; X; L	$\frac{P}{2EI}(Lx - x^2)$	At end	$\frac{P}{6EI}(3Lx^2 - 4x^3)$	At load	$\frac{P}{24EI}(3L^2x - 4x^3)$	At center
L; Total uniform load = W; X	$\frac{WL^2}{24EI}$	At end	$\frac{Wx}{24EIL}(L^3 - 2Lx^2 + x^3)$	At x from reaction	$\frac{5WL^3}{384EI}$	At center
X; Total load = W; L; L; R	$\Theta_L = \frac{7WL^2}{180EI}$ $\Theta_R = \frac{8WL^2}{180EI}$	At left reaction At right reaction	$\frac{Wx}{180EIL^2}(3x^4 - 10L^2x^2 + 7L^4)$	At x from reaction	$0.01304\,\frac{WL^3}{EI}$	At x = 0.5193L
X; Total load = W; L/2; L/2	$\frac{5WL^2}{96EI}$	At reaction	$\frac{Wx}{480EIL^2}(5L^2 - 4x^2)^2$	At x from reaction	$\frac{WL^3}{60EI}$	At center
X; M; M; L	$\frac{ML}{2EI}$	At reaction	$\frac{M}{2EI}(Lx - x^2)$	At x from reaction	$\frac{ML^2}{8EI}$	At center

x, L, M, L, R	$\Theta_L = \frac{ML}{6EI}$ $\Theta_R = \frac{ML}{3EI}$	At left reaction At right reaction	$\frac{Mx}{6EIL}(x^2 - L^2)$	At x from L reaction	$\frac{ML^2}{9(3)^{1/2}EI}$	At $x = \frac{L}{(3)^{1/2}}$
P, L, M, x	$\frac{PL^2}{2EI}$	At free end	$\frac{P}{6EI}(2L^3 - 3L^2x + x^3)$	At x from free end	$\frac{PL^3}{3EI}$	At free end
x, P, M, L	$\frac{P}{2EI}(L^2 - 2Lx + x^2)$	At load and free end	$\frac{P}{3EI}(L^3 - 3L^2x + 3Lx^2 - x^3)$	At load	$\frac{P}{6EI}(x^3 - 3L^2x + 2L^3)$	At free end
x, Total uniform load = W, L	$\frac{WL^2}{6EI}$	At free end	$\frac{W}{24EIL}(x^4 - 4L^3x + 3L^4)$	At x from free end	$\frac{WL^3}{8EI}$	At free end
x, Total load = W, L	$\frac{WL^2}{12EI}$	At free end	$\frac{W}{60EIL^2}(x^5 - 5L^4x + 4L^5)$	At x from free end	$\frac{WL^3}{15EI}$	At free end
x, Total load = W, L	$\frac{WL^2}{4EI}$	At free end	$\frac{W}{60EIL^2}(11L^5 - 15Lx^4 + 4x^5)$	At x from free end	$\frac{11WL^3}{60EI}$	At free end

Similarly, when any of the deflection equations listed in Table 6-1 are used, replace the magnitudes of x and/or L in *foot* units, and then multiply the results by 1,728 in order to obtain the correct units of y or Δ.

A timber beam 4 in. wide × 12 in. deep × 10 ft long is supported at its ends. A midpoint load of 1,800 lb is applied to the beam. $E = 1.2 \times 10^6$ psi. Determine (*a*) the maximum bending stress, (*b*) the maximum deflection, and (*c*) the slope of the beam at either reaction.

(*a*)
$$S = \frac{Mc}{I} = \frac{54{,}000(6)(12)}{4(12)^3} = 563 \text{ psi}$$

The stress is less than the allowable stress of 1,000 psi, and the design is safe.

(*b*)
$$\Delta = \frac{PL^3}{48EI} = \frac{1{,}800(10)^3}{48(1.2)10^6(4)(12)^2}(1{,}728) = 0.0938 \text{ in.}$$

This is less than the allowable deflection of $\frac{1}{360}$ of 120 in. = 0.333 in., and the design is acceptable.

(*c*)
$$\Theta = \frac{PL^2}{16EI} = \frac{1{,}800(10)^2}{16(1.2)10^6(4)(12)^2}(144) = 0.002343$$

This is approximately equal to an angle of 8 minutes or 0.134°.

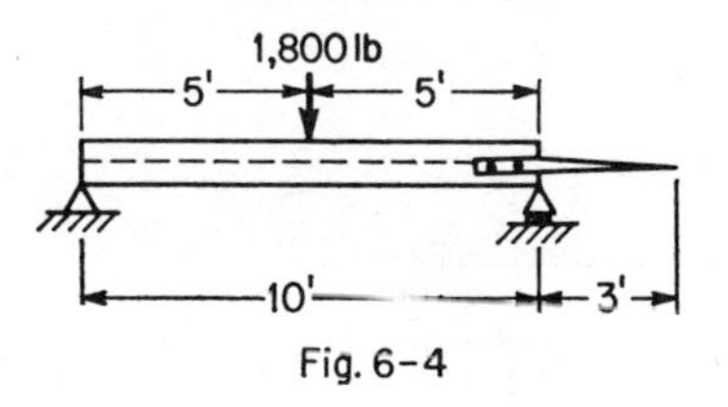

Fig. 6-4

Figure 6-4 shows the same beam and loading as were used in the preceding example. A pointer 36 in. long is rigidly attached to the neutral axis of the beam at the right support. What is the vertical displacement of the free end of the pointer after the beam is loaded?

Again, the approximation that, for small angles, the $\tan \Theta = \Theta$ and the $\cos \Theta = 1$ is used.

Letting y = the vertical displacement desired,

$$\tan \Theta = \frac{y}{36}$$

and $$y = 36 \tan \Theta = 36(0.002343) = 0.0843 \text{ in.}$$

The preceding answers emphasize the fact that the numerical magnitudes of deflections and slopes are small compared to the length of the beam.

The designer is also reminded of the fact that all deflections or vertical displacements are measured *perpendicular* to the original straight-line position of the beam. It is customary to use the tan Θ rather than the sin Θ in a calculation involving the slope Θ such as occurs in the above example.

Continuing with the same given beam and loading, what total uniform load W may be superimposed on the beam without exceeding the maximum allowable bending stress of 1,000 psi?

The net stress which can be developed by the uniform loading is

$$S = 1{,}000 - 563 = 437 \text{ psi}$$

The maximum moment for the uniform load is $WL/8$. Then

$$M = \frac{SI}{c} \qquad \frac{WL}{8} = \frac{437(4)(12)^3}{6(12)}$$

where L is in inches. Solving,

$$W = \frac{437(4)(12)^3(8)}{6(12)(120)} = 2{,}800 \text{ lb}$$

What total uniform load W may be superimposed on the beam without exceeding the maximum allowable deflection of $\frac{1}{360}$ of the span?

Both maximum deflections occur at the midpoint of the beam. Thus the net deflection caused by the uniform load will be

$$\Delta = 0.3333 - 0.0938 = 0.2395 \text{ in.}$$

From Table 6-1, for this case,

$$\Delta = \frac{5WL^3}{384EI}$$

Then

$$0.2395 = \frac{5W(10)^3}{384(1.2)(10)^6(4)(12)^2} \times 1{,}728$$

Solving for W

$$W = 7{,}300 \text{ lb}$$

If the limiting value of the superimposed total uniform load is $W = 2{,}800$ lb, what is the slope of the elastic curve at either support?

From Table 6-1

$$\theta = \frac{WL^2}{24EI} = \frac{2{,}800(10)^2}{24(1.2)(10)^6(4)(12)^2}(144) = 0.002431$$

At the supports, the total angle of slope is

$$\theta = 0.002431 + 0.002343 = 0.004774$$

and $$\theta = 16.4 \text{ minutes} \quad \text{or} \quad 0.273°$$

6-4. Bending Strain Energy. In Chaps. 1 and 3, it was possible to determine the total strain energy, and the unit strain energy, due to axial loads and torques.

Without becoming involved in the use of mathematical reason beyond the scope of this text, it is possible to determine the total strain energy due to concentrated loads. In Chap. 1, the total strain energy due to an axial load is

$$U_t = \tfrac{1}{2}Pe$$

where e is total axial deformation. By the same theory, the total strain energy for a concentrated load applied to a beam is

$$U_t = \tfrac{1}{2}Py$$

where y = deflection of the beam at the load P. When the load is applied at the point on the elastic curve where $y = \Delta$, then

$$U_t = \tfrac{1}{2}P\Delta$$

In the first numerical example in Art. 6-3, a simple beam was subjected to a concentrated load $P = 1{,}800$ lb at the midpoint of the beam. Δ occurs at the position of the load or at the midpoint of the beam. From this example, $\Delta = 0.0938$ in. Then

$$U_t = \tfrac{1}{2}(1{,}800)(0.0938)$$
$$= 84.4 \text{ in.-lb}$$

It was also noted in Chap. 1 that the unit strain energy

$$U = \frac{U_t}{\text{volume}}$$

Then the strain energy per cu in. of the material is

$$U = \frac{84.4}{4(12)(120)} = 0.0146 \text{ in.-lb/cu in.}$$

With reference to Fig. 6-3, the total strain energy will be equal to the sum of the strain energies due to P_1, P_2, and P_3 or

$$U_t = \tfrac{1}{2}P_2(\Delta_1 + \Delta_2) + \tfrac{1}{2}P_1(y_2 + y_4) + \tfrac{1}{2}P_3(y_3 + y_5)$$

and with $P_1 = P_3$,

$$U_t = \tfrac{1}{2}P_2(\Delta_1 + \Delta_2) + \tfrac{1}{2}(2)P_1(y_2 + y_4)$$

As before, the unit strain energy is the total strain energy divided by the total volume of the beam.

Problems

6-1. A cantilever timber beam 4 in. wide × 12 in. high × 10 ft long supports a concentrated load P at the free end. If the maximum allowable bending stress is 1,000 psi, what is the maximum deflection at the free end? $E = 1.2(10)^6$ psi.

6-2. A cantilever timber beam 4 in. wide × 12 in. high × 10 ft long supports a total uniform load W. $E = 1.2(10)^6$ psi. If the maximum allowable deflection is $\frac{1}{180}$ of the length, determine the maximum allowable uniform load per foot of beam.

6-3. A cantilever beam supports a concentrated load P at the free end. Determine an equation in terms of P, L, E, and I for the deflection at a distance of $L/2$ from the free end.

6-4. A cantilever beam supports a concentrated load P at the midpoint. Determine an equation in terms of P, L, E, and I for the maximum deflection.

6-5. In Prob. 6-3, when $P = 10$ kips and $L = 10$ ft, determine an equation for the radius of curvature of the beam at the midpoint in terms of E and I. Determine a similar equation for the minimum radius of curvature.

6-6. A band saw ½ in. wide × 0.05 in. thick is used as shown in Fig. 6-5. $E = 30 \times 10^6$. What force P can be applied to create tension in the blade if the maximum stress in the blade dare not exceed 50,000 psi?

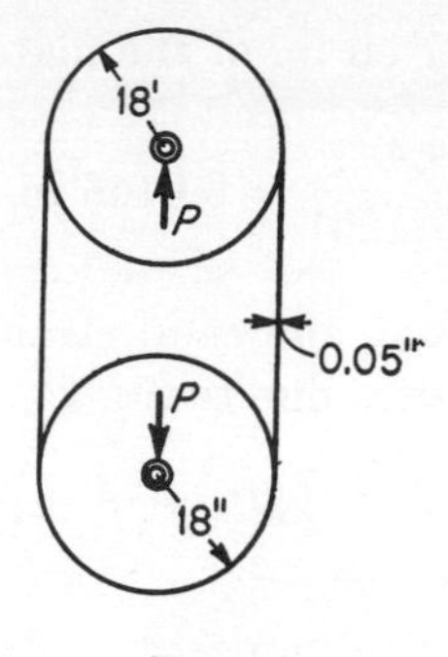

Fig. 6-5

6-7. Using the cantilever beam with two forces shown in Prob. 4-17, determine an expression for the maximum deflection in terms of P, E, and I.

6-8. A simple beam is 12 ft long and supports a concentrated load P at the midpoint. The beam is a 14-in. WF steel beam with $I = 840$ in.^{4} What is the maximum deflection in the beam? The maximum allowable stress = 24,000 psi.

6-9. In Prob. 6-8, what is the radius of curvature at a point on the elastic curve 3 ft from the left reaction?

6-10. Using the 14-in. WF beam of Prob. 6-8, what total uniform load W will the beam safely support? What is the maximum deflection in the beam? If the maximum deflection is limited to $\frac{1}{360}$ of the span, does the beam satisfy this specification? Use $S = 24,000$ psi.

6-11. Using the 14-in. WF beam of Prob. 6-8 and referring to Fig. 6-6, determine the magnitude of each of the three equal loads. Use $S = 24,000$ psi.

6-12. What is the maximum deflection in the 14-in. WF beam referred to in Prob. 6-11?

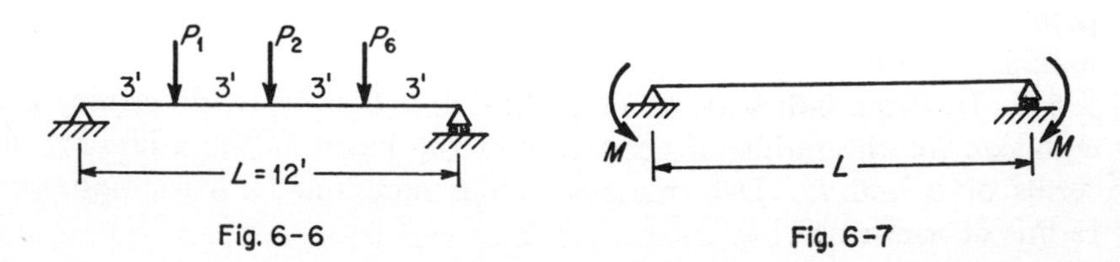

Fig. 6-6

Fig. 6-7

6-13. A simple beam with a span of 22 ft is to support a single concentrated load $P = 40$ kips at the midpoint. The beam is limited to an 18-in. WF section by specification. What is the required size (weight per foot) of the steel beam if the stress must not exceed 22,000 psi and the deflection must not exceed $\frac{1}{360}$ of the span?

6-14. The simple steel beam shown in Fig. 6-7 is 15 ft long with $M = 300$ ft-kips. The beam is a 24-in. WF at 84 lb. $E = 30 \times 10^6$ psi. What is the maximum displacement of the elastic curve? Is this displacement above or below the level of the supports?

6-15. Figure 6-8 shows the same beam as in Fig. 6-7 with the beam extended by 4 ft on each end. The beam is assumed to be weightless, and a 11.5-ft weightless pointer is rigidly attached to the beam at the midpoint of the elastic curve. Using a 24-in. WF-at-84-lb beam, what is the vertical distance from the free end of the pointer to the elastic curve when the 300-ft-kips equal moments are applied at the supports?

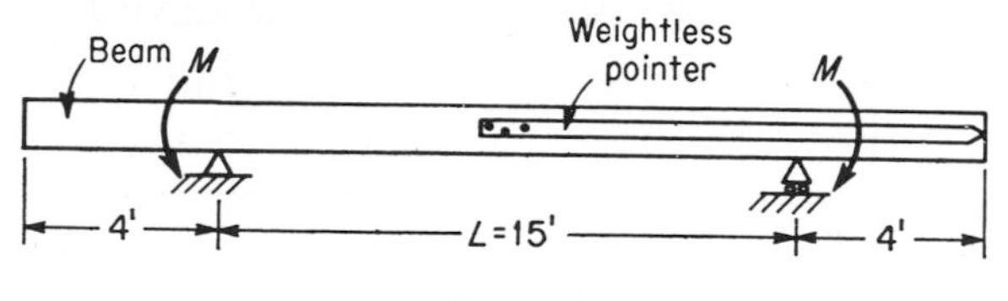

Fig. 6-8

6-16. A WF beam section is used as a simple beam 15 ft long and is loaded as shown in Fig. 6-9. The total nonuniform load is 60 kips. If the maximum deflection is limited to $\frac{1}{360}$ of the span, what is the required minimum moment of inertia of the beam? What size beam would you recommend if the depth of the beam may not exceed 12 in.?

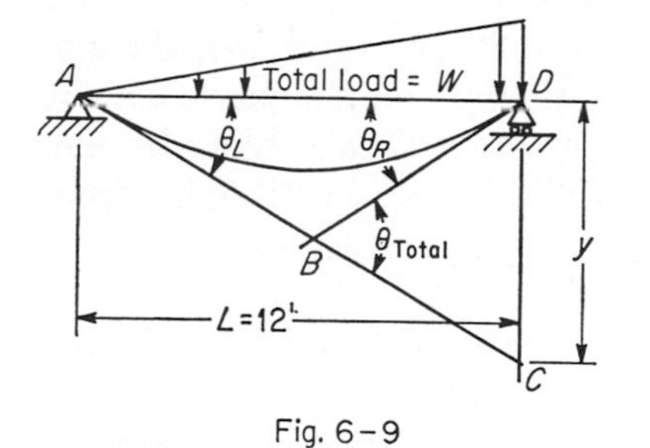

Fig. 6-9

6-17. A timber cantilever beam is 10 ft long with a width b and a depth $2b$. The beam is subjected to a total uniform load of 12,000 lb. $E = 1.2 \times 10^6$ psi. If the maximum slope of the beam is limited to $\Theta = 0.120$, what is the minimum width of beam?

6-18. A simple steel beam 12 ft long is subject to a total nonuniform loading of 30 kips as shown in Fig. 6-9. $I = 2{,}400$ in.^{4} The line ABC is tangent to the elastic curve at A. Line BD is tangent to the elastic curve at D. Determine the total angle Θ_T.

6-19. In Prob. 6-18, determine the distance y from D to C shown in Fig. 6-9.

6-20. A simple beam of length L is loaded by a total uniform load W and two symmetrically placed equal loads, each $= P$. The P loads are placed at the quarter points of the beam. The load $W = 2P$. Determine the deflection of the beam at the quarter point or at $L/4$ from either reaction, in terms of P, L, E, and I.

6-21. The following data apply to the simple beam shown in Fig. 6-10: $P = 10{,}000$ lb; $L = 15$ ft; $I = 2{,}400$ in.^{4}; $E = 1.2 \times 10^6$ psi. The pointer is rigidly attached to the end of the beam at the elastic curve. Determine the vertical distance between the free end of the pointer and the elastic curve after the load P is applied.

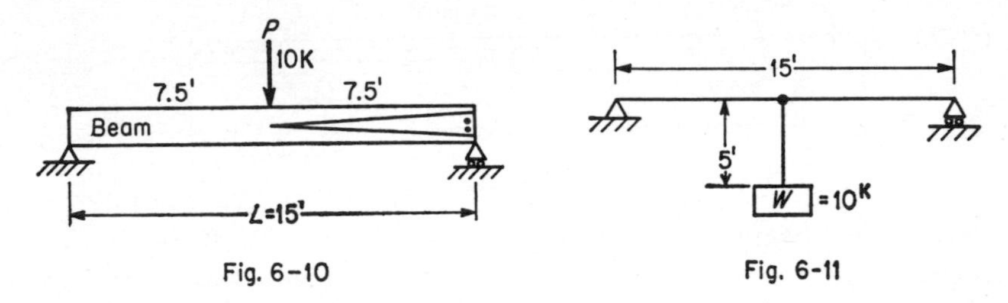

Fig. 6-10

Fig. 6-11

6-22. An aluminum cantilever beam is 9 ft long with $I = 100$ in.^{4} It supports two equal loads P and P. One load is applied at the free end, and the other acts at 3 ft from the free end. If the maximum deflection is limited to 0.60 in., determine the maximum load P.

6-23. A steel I beam 15 ft long is used as a simple beam 15 ft long (Fig. 6-11). A weight of 10 kips is hung by means of an aluminum alloy bar 5 ft long, with an area $= 0.50$ sq in., from the midpoint of the beam. Determine the total vertical displacement of the weight. For the beam, $I = 100$ in.^{4}

6-24. Referring to Prob. 6-1, determine the total elastic strain energy in the deflected position of the beam.

6-25. Referring to Prob. 6-8, determine the unit elastic strain energy in the 14-in. WF beam after the load P has been applied. Use the area $A = 23.0$ sq in.

6-26. A 12-in. WF steel beam has $I = 840$ in.^{4} It is used as shown in Fig. 6-12 where the loads $P = P = 40{,}000$ lb. Determine: (*a*) the deflection at either of the symmetrically positioned loads; (*b*) the total strain energy in the beam.

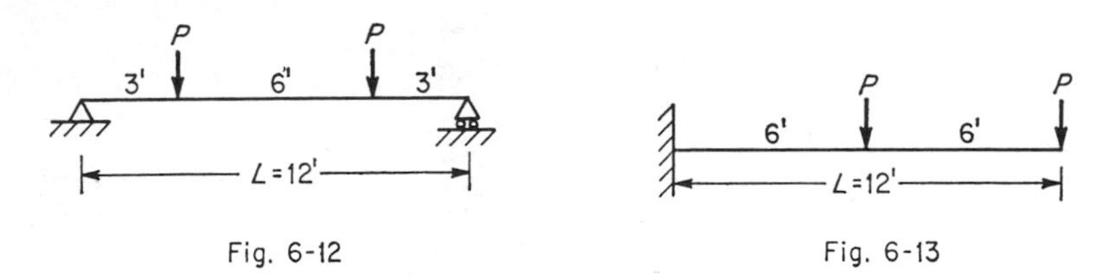

Fig. 6-12

Fig. 6-13

6-27. In Fig. 6-13, $P = P = 10$ kips. The beam is steel with $I = 2(576)$ in.^{4} Determine the maximum deflection.

6-28. In Fig. 6-13, $P = P = 10$ kips. The beam is steel with $I = 2(576)$ in.^{4} Determine the total vertical displacement of the elastic curve at the midpoint load.

6-29. Referring to Probs. 6-27 and 6-28, determine the total strain energy absorbed by the beam when it is loaded.

6-30. Figure 6-6 shows a simple steel beam supporting three equal concentrated 40,000-lb loads. When the beam is a 14-in. WF with $I = 2(576)$ in.^{4}, determine: (*a*) the deflection at P_1 or P_3 due to the *three* loads on the beam; (*b*) the total maximum deflection at the center line; (*c*) the total strain energy in the beam.

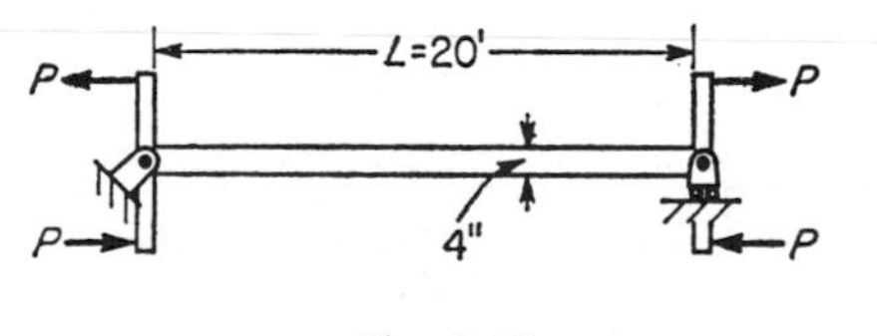

Fig. 6-14

6-31. A 3-in.-wide × 4-in.-high × 20-ft-long piece of aluminum alloy is to be used as a beam. Opposing couples, each with a value of 80,000 in.-lb, act at the ends of the beam shown in Fig. 6-14. Using $E = 10 \times 10^6$ psi, determine the change in length of the top (or bottom) layer of the beam after the couples have been applied.

CHAPTER 7

COMBINED AXIAL AND BENDING STRESSES—ECCENTRIC LOADS ON RIVETED CONNECTIONS

7-1. Axial and Bending Stresses. In many cases of engineering practice and design, the stresses which are developed in a piece of deformable material do not act singly. In some cases, they are superimposed so that they act in conjunction with each other; in other cases, they oppose each other. As long as the axial and bending stresses act collinear, they are addable algebraically. It is customary to consider

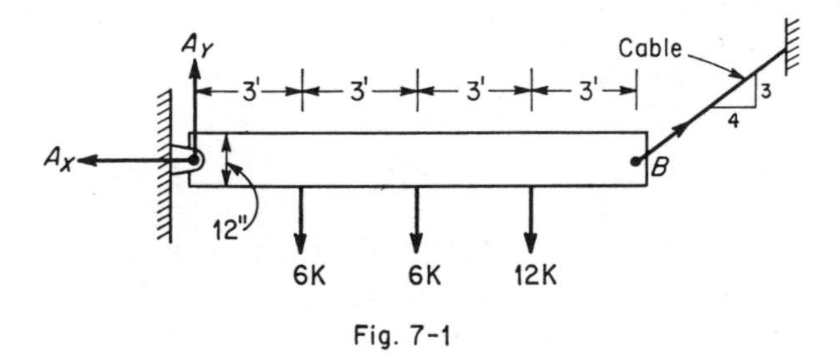

Fig. 7-1

tensile stresses as being positive; hence, compressive stresses are considered algebraically negative. By the principle of superposition, any number of collinear stresses can be combined to determine the resultant stress which is acting at a given point or on a given particle of the material.

An excellent example of combined axial and bending stresses comes from the classic statics problem shown in Fig. 7-1. The beam, which is 2 in. wide and 12 in. high, is pinned at A and supported at B by the cable. Analyzing the problem, step by step, to determine the maximum tensile and compressive stresses in the beam, we obtain the following results:

Solving the components of the pin reaction at A and the cable tension

$$\Sigma M_A = 0 = -3'(6) - 6'(6) - 9'(12) + 12(\tfrac{3}{5}B)$$

gives $\quad B = 22.5 \text{ kips}$

$$\Sigma M_B = 0 = -3'(12) - 6'(6) - 9'(6) + 12A_y$$

gives $\quad A_y = 10.5 \text{ kips}$

Checking: $\quad \Sigma F_y = 0 = \tfrac{3}{5}(22.5) + 10.5 - 6 - 6 - 12$

By $\quad \Sigma F_x = 0 = \tfrac{4}{5}(22.5) - A_x$

$$A_x = 18 \text{ kips}$$

The vertical forces acting on the (simple) beam create bending stresses. A shear diagram shows that the maximum moment in the beam occurs at the section of the beam where the middle load is applied. Then

$$M_{\max} = +6'(10.5) - 3'(6) = +45 \text{ ft-kips}$$

The maximum bending stress S is

$$S = \frac{Mc}{I} = \frac{45{,}000(12)6}{2(12^3)/12} = 11{,}250 \text{ psi}$$

This stress is either tensile stress at the bottom or compressive stress at the top of the beam. A sketch of the elastic curve indicates the correct positioning of the type of stresses.

The horizontal forces A_x and $\tfrac{4}{5}B$ create an axial tensile stress condition from which the axial stress S is

$$S = \frac{P}{A} = \frac{18{,}000}{2(12)} = 750 \text{ psi tension}$$

Recalling that the bending stresses on a beam cross section are horizontal, as are the axial stresses, then

$$S_{\max} = +11{,}250 + 750 = 12{,}000 \text{ psi tension at bottom of beam}$$

and $\quad S_{\min} = -11{,}250 + 750 = -10{,}500 \text{ psi compression}$

From the preceding example, we can conclude that a very general equation can be formulated as

$$\left.\begin{matrix} S_{\max} \\ \text{or} \\ S_{\min} \end{matrix}\right\} = \pm \frac{P}{A} \pm \frac{Mc}{I} \pm \left[\frac{Mc}{I}\right]$$

The last term has been added to indicate that more than one bending moment may be acting on the member which is being stress-analyzed.

A somewhat different analysis must be made of the applied force system shown in Fig. 7-2*a*, *b*, and *c*. In this situation, only one vertical force P is applied to the corner D of the short block. It will be assumed that the block acts only as a direct compression member. In all parts, the force

$$P_1 = P_1' = P_2 = P_2' = P_3 = P_3' = P$$

When a force P_1 is applied at F in Fig. 7-2*a*, it can be replaced by a force *and* a couple by adding two equal, opposite, collinear forces P_1'

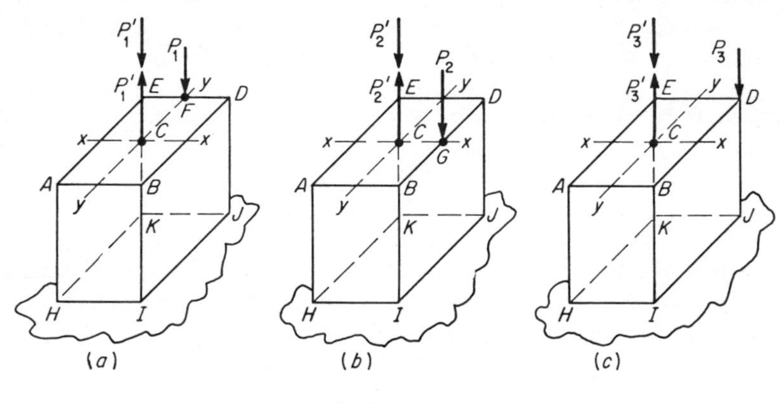

Fig. 7-2

and P_1' at the centroid C of the top area. The couple causes the top of the block to bend backward, with the X-X axis becoming the neutral axis. The moment of the couple is the moment M_1 which sets up the internal resisting tensile stresses along HI and compressive stresses along JK. The other force P_1' causes axial compressive stresses to act *uniformly* over the base area.

When P_2 is applied at G in Fig. 7-2*b*, it can be replaced by a force and a couple by adding two equal, opposite, collinear forces P_2' and

P'_2 at the centroid C. This couple causes the top of the block to bend to the right, with the Y-Y axis becoming the neutral axis. The moment of the couple is the moment M_2 which results in tensile resisting stresses along HK and compressive resisting stresses along IJ. The other force P'_2 acts through the centroid of the top area, resulting in a uniform axial compressive stress over the base area $HIJK$.

If the force $P = P_3$ is now shifted to the corner D, it can likewise be replaced by a force and a couple by adding two equal, opposite, collinear forces at C. This couple is the *equivalent* of the two couples described in the preceding two paragraphs and produces the same effects as both of the P'_1 and P'_2 couples. The single force P'_3 acting downward at C causes the same uniform compressive stress as *either* of the single forces P'_1 or P'_2 which was an axial compressive force.

Note. The $P'_1\ \overline{CF}$ couple plus the $P'_2\ \overline{CG}$ couple equals the $P'_3\ \overline{CD}$ couple. But note that only one axial compressive force $P'_1 = P'_2 = P'_3$ acts. (It can be shown that any couple can be represented by a vector, and the vector representing the $P'_3\ \overline{CD}$ couple is the resultant of the vectors representing the other two couples. Reversing the last statement, the $P'_1\ \overline{CF}$ vector and the $P'_2\ \overline{CG}$ vector are the X and Y components of the $P_3\ \overline{CD}$ vector.)

Assuming the following numerical data for Fig. 7-2, let $HI = 4$ in.; $IJ = 6$ in.; $P = 4{,}800$ lb at D. The bending stress from the $P_1\ \overline{CF}$ couple is

$$S = \frac{Mc}{I} = \frac{4{,}800(3)3}{4(6)^3/12} = 600 \text{ psi}$$

which is tension at HI and compression at JK.

The bending stress from the $P_2\ \overline{CG}$ couple is

$$S = \frac{Mc}{I} = \frac{4{,}800(2)2}{6(4)^3/12} = 600 \text{ psi}$$

which is tension at HK and compression at IJ.

The axial compressive stress due to the applied force P is

$$S = \frac{P}{A} = \frac{4{,}800}{24} = 200 \text{ psi}$$

which is uniformly compression on the base area $HIJK$.

Adding the stresses at each of the corners at the base:

At H: $\quad S = +600 + 600 - 200 = +1{,}000$ psi

At I: $\quad S = +600 - 600 - 200 = -200$ psi

At J: $\quad S = -600 - 600 - 200 = -1{,}400$ psi

At K: $\quad S = -600 + 600 - 200 = -200$ psi

Plotting these stresses at the four corners, as seen in Fig. 7-3, and recalling that the stresses must vary linearly as their distances from the true neutral axis, we can locate the points L and N on the lines along which the stress must be equal to zero.

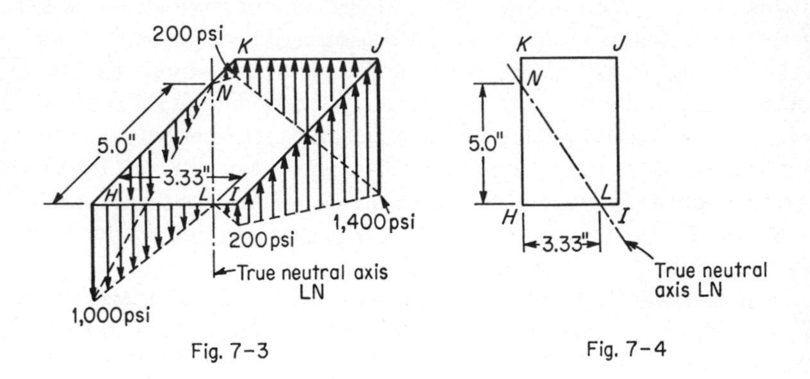

Fig. 7-3 Fig. 7-4

Along the line HI, the stress at $H = +1{,}000$ psi and at $J = -200$ psi. $HI = 4$ in. By proportion

$$\frac{HL}{1{,}000} = \frac{LJ}{200} \qquad \frac{HL}{1{,}000} = \frac{4 - HL}{200}$$

from which $HL = 3.33$ in.

Along the line KH, the stress at $K = -200$ psi and at $H = +1{,}000$ psi. By proportion

$$\frac{HN}{1{,}000} = \frac{NK}{200} \qquad \frac{HN}{1{,}000} = \frac{6 - HN}{200}$$

from which $HN = 5.0$ in.

A plan of the base area is given in Fig. 7-4, on which the true neutral axis is shown. This can be used to interpret the true direction of bending which is perpendicular to the true neutral axis.

Certain other facts should be observed:

1. The maximum compressive stress occurs at the corner where the load P was applied.

2. The opposite corner must be the corner where the maximum opposite type of stress occurs. (At corner J, $S = -1{,}400$ psi, and at corner H, $S = +1{,}000$ psi.)

3. The maximum shortening of the block occurs along DJ.

4. The maximum stretch occurs along the opposite edge AH.

The use of stress diagrams such as were shown in Fig. 7-2 is helpful in visualizing the stress conditions which exist.

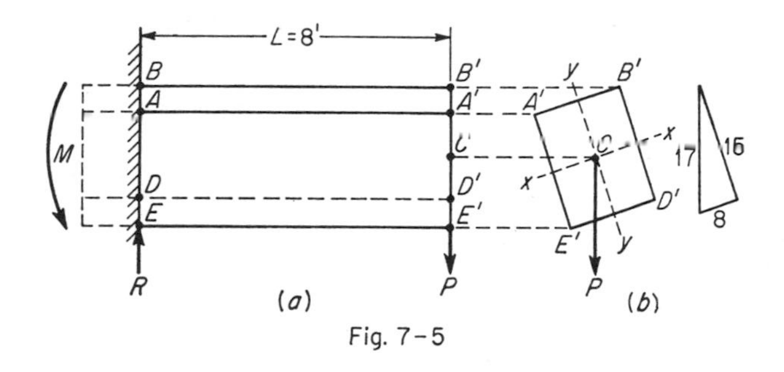

Fig. 7-5

7-2. Bending When the Load Is Not in a Principal Plane. Figure 7-5*a* and *b* represents a cantilever beam of length $L = 8$ ft with a rectangular cross section with $A'B' = 4$ in. and $B'D' = 6$ in. A *vertical* load $P = 510$ lb is applied to the centroid C of the cross section. The stresses at each of the corners A, B, D, and E of the wall cross section of the beam are required. Shearing stresses at each of the corners noted are small and will be neglected.

The analysis of the stress conditions due to the bending of the beam is simplified by analyzing the two directions of simultaneous bending. By resolving the load P into a component perpendicular to the X-X axis in Fig. 7-5*b*, and a component perpendicular to the Y-Y axis, their respective bending stresses can be superimposed at the four corners A, B, D, and E.

Noting the $15 \nrightarrow 8 = 17$ right triangle shown in Fig. 7-5*b*, we find the component of P in the direction of the Y-Y axis is

$$510(^{15}/_{17}) = 450 \text{ lb down to the right}$$

and in the direction of the X-X axis is

$$510(^{8}/_{17}) = 240 \text{ lb down to the left}$$

The neutral axis for the 450-lb component is the X-X axis; for the 240-lb component, the neutral axis is the Y-Y axis.

For the X-X bending

$$S = \frac{Mc}{I} = \frac{450(96)3}{4(6)^3/12} = 1{,}800 \text{ psi}$$

This stress is tension on the top edge AB and is compression on the bottom edge DE.

For the Y-Y bending

$$S = \frac{Mc}{I} = \frac{240(96)2}{6(4)^3/12} = 1{,}440 \text{ psi}$$

This stress is tension on the back edge BD and is compression on the front edge AE.

Adding the stresses,

At A: $S = +1{,}800 - 1{,}440 = +360$ psi

At B: $S = +1{,}800 + 1{,}440 = +3{,}240$ psi

At D: $S = -1{,}800 + 1{,}440 = -360$ psi

At E: $S = -1{,}800 - 1{,}440 = -3{,}240$ psi

An examination of the resulting stresses indicates the expected maximums; that is, the edge $B'B$ will have the maximum elongation, and edge $E'E$ will have the maximum shortening.

With the same method which was used for locating the true neutral axis at the base of the short block, the true neutral axis for the cantilever beam can be located.

7-3. Biaxially Loaded Riveted Connections. In Chap. 2, a complete discussion of riveted connections was given. However, in that discussion the applied load was always uniaxial and created a condition where the entire connection was analyzed for shearing, tensile, and compressive stresses in the rivets and connected plates. The line of action of the applied load always acted through the centroid of the rivet pattern.

There are other design conditions where the line of action of the applied load passes through the centroid of the rivet pattern but is not applied perpendicular to the rows of rivets. This becomes essentially a biaxial loading situation, where the load is resolvable into its components, parallel and perpendicular to the rows of rivets. Shear, only, will be considered critical in this analysis, and the other possible stresses are assumed to be safe.

Examining a biaxially loaded riveted connection, Fig. 7-6*a* and *b*, where all rivets are in single shear, we find the load $P = 10{,}400$ lb is applied to the centroid C of the riveting pattern. As shown in Fig. 7-6*b*, the load has been resolved into its X and Y components acting at

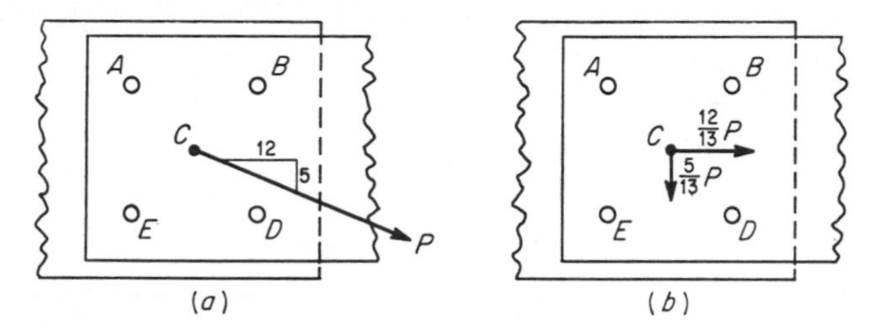

Fig. 7-6

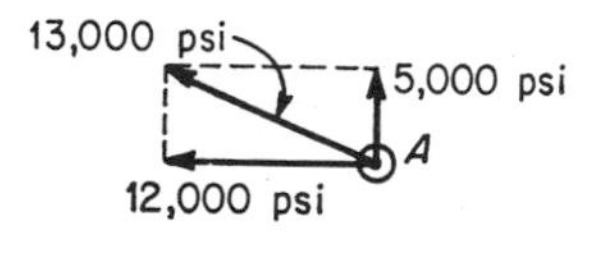

Fig. 7-7

C. Assuming that each rivet has a shear area of 0.20 sq in., the X component of the load P produces a uniform shear stress

$$S_{s_x} = \frac{P_x}{A} = \frac{9{,}600}{4(0.20)} = 12{,}000 \text{ psi}$$

which is in the X direction to the left.

For the Y component of the load P, the uniform shear stress

$$S_{s_y} = \frac{P_y}{A} = \frac{4{,}000}{4(0.20)} = 5{,}000 \text{ psi}$$

which acts upward in the Y direction.

The preceding shear stresses are in the plane of the cross-sectional areas of the rivets and can be superimposed to obtain their resultant effect. Figure 7-7 shows the X and Y shear stresses acting on rivet A, and their resultant stress $= 13{,}000$ psi. Since all rivet areas transmit equal parts of the applied *load*, the resultant shear stress in all rivets will be the same.

The same result can be obtained by

$$S_s = \frac{P}{A} = \frac{10{,}400}{4(0.20)} = 13{,}000 \text{ psi}$$

which has the *same direction* as the applied load but has the *opposite sense.* The usefulness of the component method will become more apparent as explained in the next article.

7-4. Eccentrically Loaded Riveted Connections. When a riveted connection is required to transmit a torque or moment, the analysis of

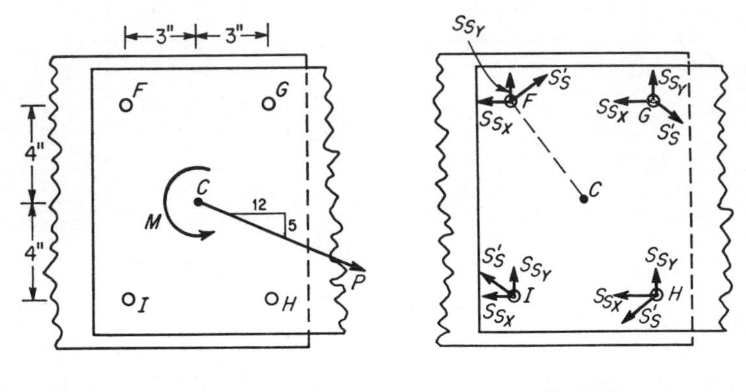

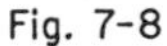
Fig. 7-8

Fig. 7-9

the shear stresses is the same as that of the shear stresses in a bolted flanged coupling. See Art. 3-6. When a moment and force P are both applied to the connection as shown in Fig. 7-8, the superimposed shear stresses provide the resultant stress on each rivet. The shear stress on all rivets will not be the same, as the following example indicates.

Assume that $M = 1{,}000$ ft-lb, $P = 7{,}800$ lb, and $A = 0.20$ sq in. per rivet.

$$S_{s_x} = \frac{\frac{12}{13}(7{,}800)}{4(0.20)} = \frac{7{,}200}{0.80} = 9{,}000 \text{ psi}$$

which is uniform on all rivets and is horizontal to the left.

$$S_{s_y} = \frac{\frac{5}{13}(7{,}800)}{4(0.20)} = \frac{3{,}000}{0.80} = 3{,}750 \text{ psi}$$

which is uniform on all rivets and is vertically upward.

$$M = T = 4(5)S_s'(0.20)$$
$$12{,}000 = 4S_s'$$
$$S_s' = 3{,}000 \text{ psi}$$

which is due to the applied moment or torque and is perpendicular to the radius from C to the center of each rivet area. The direction of S_s' on each of the rivets is shown in Fig. 7-9, as are the directions of the S_{sx} and S_{sy} shear stresses. The resultant stresses are:

On rivet F:

$$S_s = \{[+\tfrac{4}{5}(3{,}000) - 9{,}000]^2 + [+\tfrac{3}{5}(3{,}000) + 3{,}750]^2\}^{1/2}$$
$$= [(-6{,}600)^2 + (+5{,}550)^2]^{1/2} = 8{,}620 \text{ psi}$$

On rivet G:

$$S_s = \{[+\tfrac{4}{5}(3{,}000) - 9{,}000]^2 + [-\tfrac{3}{5}(3{,}000) + 3{,}750]^2\}^{1/2}$$
$$= [(-6{,}600)^2 + (+1{,}950)^2]^{1/2} = 6{,}880 \text{ psi}$$

On rivet H:

$$S_s = \{[-\tfrac{4}{5}(3{,}000) - 9{,}000]^2 + [-\tfrac{3}{5}(3{,}000) + 3{,}750]^2\}^{1/2}$$
$$= [(-11{,}400)^2 + (+1{,}950)^2]^{1/2} = 11{,}600 \text{ psi}$$

On rivet I:

$$S_s = \{[-\tfrac{4}{5}(3{,}000) - 9{,}000]^2 + [+\tfrac{3}{5}(3{,}000) + 3{,}750]^2\}^{1/2}$$
$$= [(-11{,}400)^2 + (+5{,}550)^2]^{1/2} = 12{,}700 \text{ psi}$$

From the above four values of S_s, the maximum shear stress occurs on rivet I.

A similar loading system on a riveted connection occurs when an eccentric (to the centroid of the rivet pattern) load is applied. Figure 7-10 shows a single eccentric load $P = 6{,}000$ lb applied to the rivet pattern consisting of four rivets in a line. $A = 0.20$ sq in. per rivet.

The eccentric load P can be replaced by a force and a couple by the addition of two equal and opposite forces P' and P' (also shown) acting through C. The loading system is now composed of an equivalent system consisting of the clockwise moment or couple *and* a single force

$P' = P$ acting at C. Considering the uniform shear stress in each rivet due to the concentric force at C,

$$S_{s_y} = \frac{6{,}000}{4(0.20)} = 7{,}500 \text{ psi downward}$$

Considering the couple,

$$M = 6{,}000(7/2) = 21{,}000 \text{ in.-lb}$$

Recalling that the stress on any rivet in a group of rivets, being acted

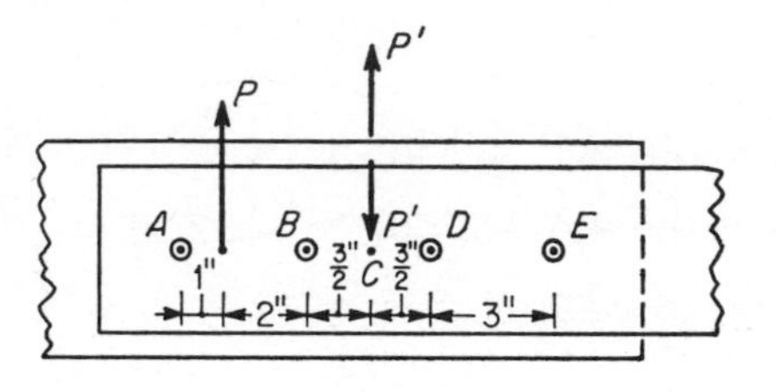

Fig. 7-10

upon by a torque or moment, varies directly as the distance from the centroid of the rivet pattern,

$$S'_{s_A} : 9/2 :: S'_{s_B} : 3/2 :: S'_{s_D} : 3/2 :: S'_{s_E} : 9/2$$

from which $\qquad S'_{s_A} = S'_{s_E} = 3S'_{s_B} = 3S'_{s_D}$

Then $\qquad M = (0.20)\,9/2\,S'_{s_A} + (0.20)\,3/2\,S'_{s_B} + (0.20)\,3/2\,S'_{s_D} + (0.20)\,9/2\,S'_{s_E}$

Noting that $S'_{s_A} = S'_{s_E}$ and replacing S'_{s_B} and S'_{s_D} by

$$S'_{s_B} = S'_{s_D} = 1/3\,S'_{s_A}$$

$$21{,}000 = 2(0.20)\,9/2\,S'_{s_A} + 2(0.20)\,3/2\,(1/3)S'_{s_A}$$

$$= 2.0S'_{s_A}$$

from which $\qquad S'_{s_A} = 10{,}500 \text{ psi}$

The S'_s stress must rotate counterclockwise about C to resist the clockwise couple. Then S'_{s_A} and S'_{s_B} will act vertically downward, while

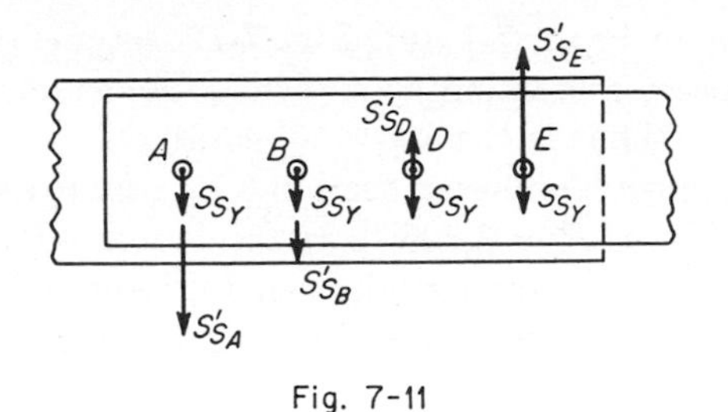

Fig. 7-11

S'_{s_D} and S'_{s_E} must act vertically upward. The component stress system is shown in Fig. 7-11. The resulting stresses are:

On rivet A: $S_s = -7{,}500 - 10{,}500 = -18{,}000$ psi down

On rivet B: $S_s = -7{,}500 - (\frac{1}{3})10{,}500 = \quad 11{,}000$ psi down

On rivet D: $S_s = -7{,}500 + \frac{1}{3}(10{,}500) = -4{,}000$ psi down

On rivet E: $S_s = -7{,}500 + 10{,}500 = +3{,}000$ psi up

The plus or minus signs with the answers are conventional signs. They are not an integral part of the values. The maximum stress occurs in rivet A as will be noted.

Problems

7-1. A cantilever beam 2 in. wide $\times$ 12 in. high $\times$ 4 ft long has three forces P_1, P_2, and P_3 acting on it. P_1 and P_2 act at the centroid of the end cross section, and the line of action of P_3 passes through the centroid of its cross section. Letting $P_1 = 24$ kips, $P_2 = 1$ kip, and $P_3 = 2$ kips, determine the stresses at points A and B at the wall. See Fig. 7-12.

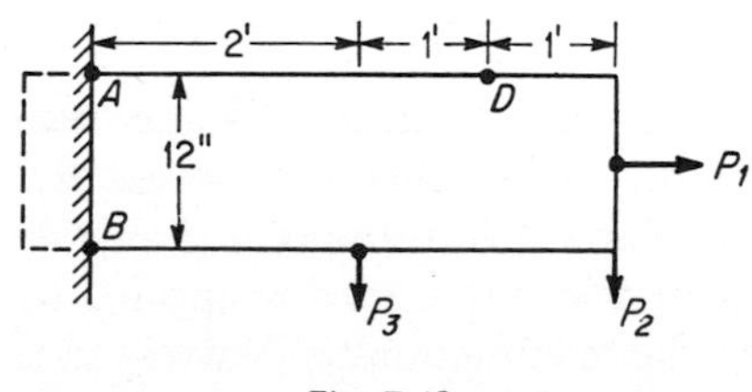

Fig. 7-12

7-2. Referring to Prob. 7-1 and Fig. 7-12, let P_2 act upward with the other two forces remaining as shown. Determine the stresses at points A and B. What is the stress at point D?

7-3. A vertical post has a cross section 6 in. square and is 12 in. high. A vertical *tensile* force $P = 3{,}600$ lb is applied on one of the principal axes at a point midway between the centroid and an edge of the top area. Determine the maximum and minimum stresses at the bottom of the post.

7-4. If the tensile force $P = 3{,}600$ lb of Prob. 7-3 is applied at a distance of $(2)^{1/2}$ in. from the centroid on a line joining the centroid and a corner of the top of the post, determine the maximum and minimum stresses at the bottom of the post.

7-5. In Prob. 7-4, how far from the centroid, on the same diagonal of the top of the post as the load was applied, does the true neutral axis cross this diagonal?

7-6. A short, vertical post is shown in Fig. 7-13. $I_{xx} = 136$ in.^{4} When a force $P = 24{,}000$ lb is applied as shown, determine the maximum and minimum stresses at the base of the post.

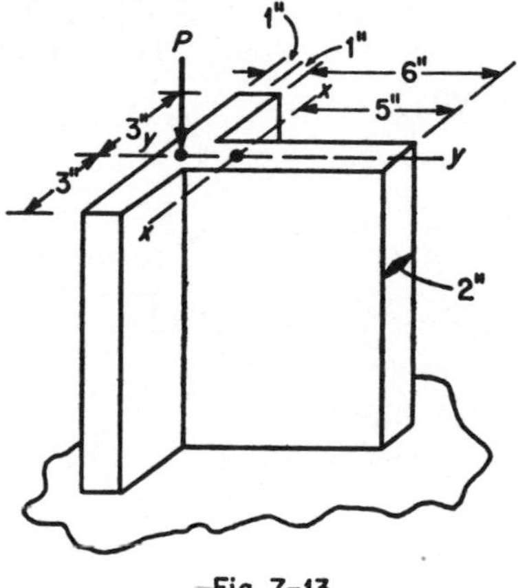

Fig. 7-13

7-7. In Prob. 7-6, at what maximum distance from the X axis can the force be applied, so that there is no tensile stress at the base of the post?

7-8. Figure 7-14 shows a 5-ft-long cantilever beam with a square cross section. The concentrated load is applied to the free end of the beam. Determine the maximum tensile stress in the cantilever beam. At which corner of the wall cross section does this stress act? At which corner at the wall does the maximum compressive stress act, and what is the magnitude of this stress?

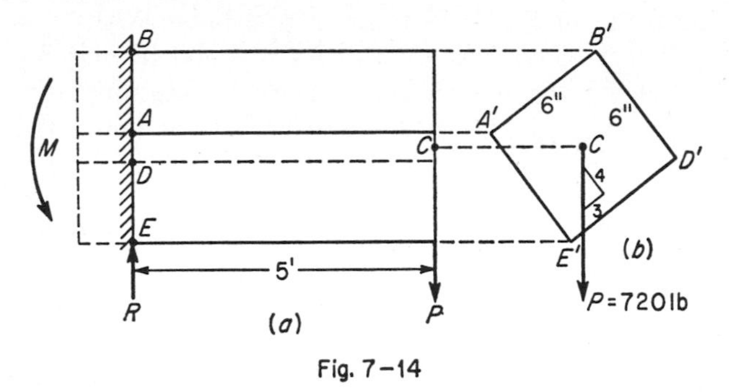

Fig. 7-14

7-9. A 2-in.-diameter bar is bent into the form shown in Fig. 7-15. When $P = 1{,}000\pi$ lb, determine the stresses which are developed at points C and D.

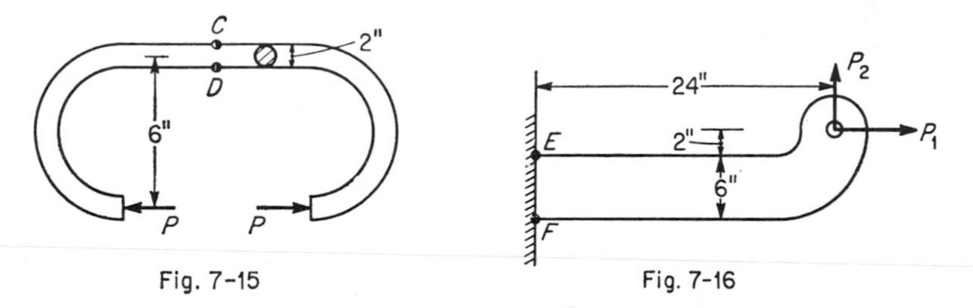

Fig. 7-15

Fig. 7-16

7-10. A steel plate, 2 in. thick, is cut in the form shown in Fig. 7-16. When $P_1 = 24{,}000$ lb and $P_2 = 0$, determine the stresses which are developed at points E and F.

7-11. In Fig. 7-16 and Prob. 7-10, what is the magnitude of P_2 when the stress at point F is zero?

7-12. A steel T beam is shown in Fig. 7-17. The centroid of the cross section is at C which is at the junction of the web and flange. $I_x = 20$ in.^{4}, and $A = 10$ sq in. When $P_3 = 5{,}000$ lb, determine the stresses at points G and H.

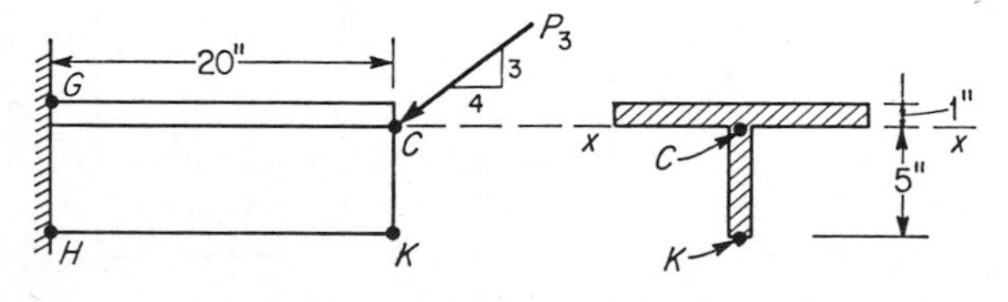

Fig. 7-17

7-13. In Fig. 7-17 and Prob. 7-12, if P_3 acts at point K, with everything else the same, determine the stresses at G and H.

7-14. The beam in Fig. 7-18 is an 8-in. I beam with $A = 9$ sq in. and $I_x = 120$ in.^{4} The forces $P_4 = P_4 = 9{,}900$ lb and $P_1 = P_2 = P_3 = 12$ kips each. Determine the numerical maximum and minimum stresses in the beam, and state the kind of each of these stresses.

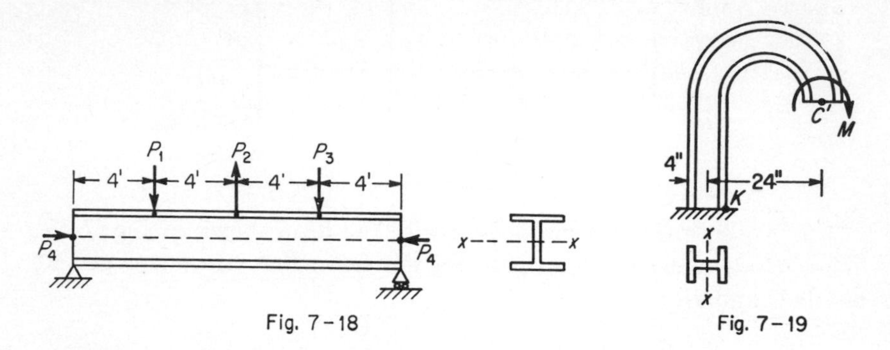

Fig. 7-18 Fig. 7-19

7-15. A moment $M = 10{,}000$ ft-lb is applied to the free end of the machine part as shown in Fig. 7-19. The cross-sectional elements are $A = 9$ sq in.; $I_x = 120$ in.^{4} What upward force P can be applied at C' so that the stress at K is zero?

7-16. The diagonal cantilever of length L has a vertical load P applied to the centroid of the end cross section shown in Fig. 7-20. For the condition of zero compressive stress in the beam, what is the maximum length L?

7-17. An eccentrically loaded triple riveted connection is shown in Fig. 7-21. The rivet areas are 0.20 sq in. each. When the load $P_1 = 3$

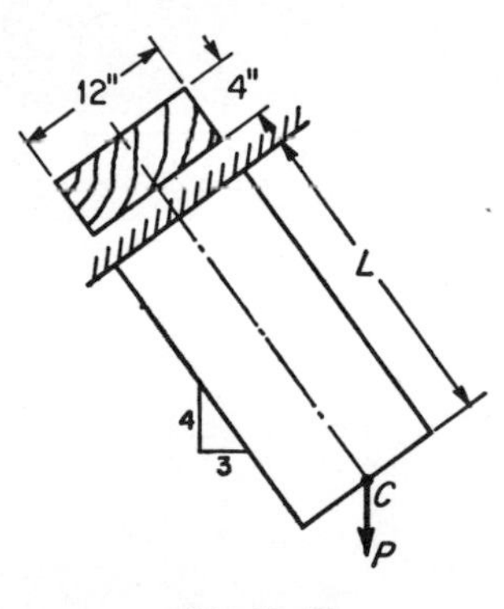

Fig. 7-20

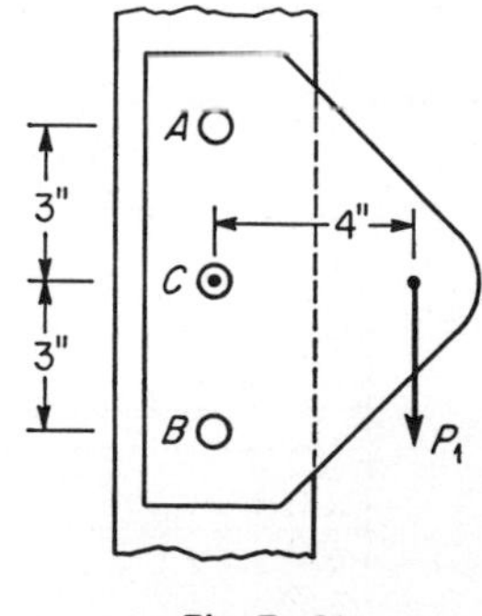

Fig..7-21

kips, what is the shearing stress in the rivet B? (To simplify the calculations, the author is using a rivet area = 0.20 sq in. This is, approximately, the area of a ½-in.-diameter rivet. The approximate area of a ¾-in.-diameter rivet is being used as 0.40 sq in.)

7-18. What additional *concentric* load P_2 can be applied to the connection shown in Fig. 7-21, and in what direction must it be applied so that the maximum stress in rivet A does not exceed 14,500 psi in shear?

7-19. A riveted connection is formed by four rivets arranged symmetrically around a 2-in.-radius circle as shown in Fig. 7-22. Each rivet area = 0.20 sq in. When P_3 = 2,400 lb, determine the shearing stress in each of the four rivets.

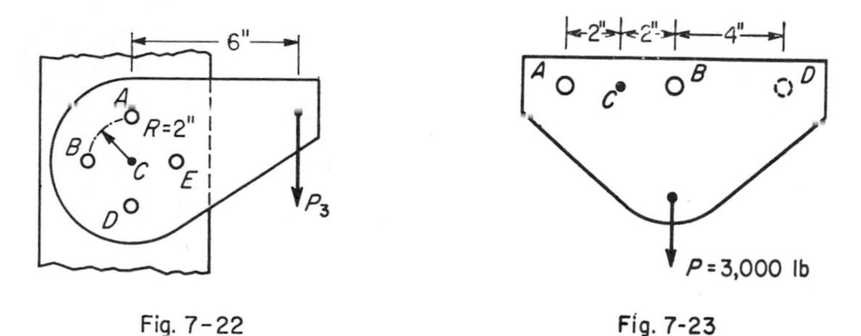

Fig. 7-22 Fig. 7-23

7-20. A riveted connection, with each rivet area = 0.40 sq in., is shown in Fig. 7-23. When the connection consists of rivets A and B, what is the shearing stress in each of the rivets?

7-21. If a third rivet were driven at D in Fig. 7-23, what is the shearing stress in each of the three rivets? What is the ratio of the maximum shearing stress in the two-rivet connection to the maximum shearing stress in the three-rivet connection?

7-22. A 12-in.-long cantilever beam has a T-shaped cross section as shown in Fig. 7-24. The area = 4 sq in. and $I_x = 6.0 \text{ in.}^4$ When a load P = 5,000 lb is applied at the centroid of the area, what unit stress is developed at points R and U? How far above the bottom edge of the beam is the true neutral axis?

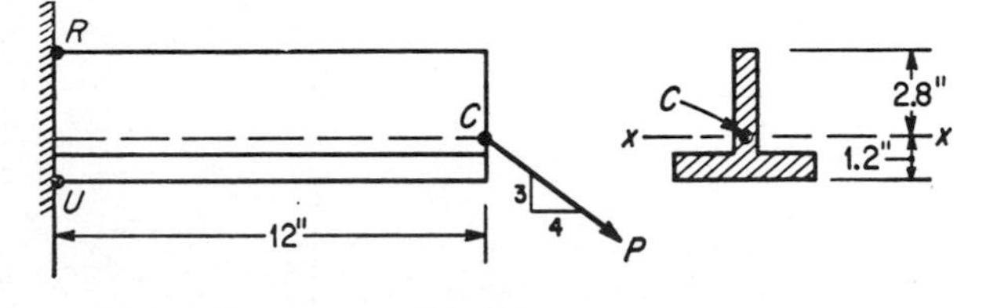

Fig. 7-24

7-23. An eccentrically loaded riveted connection has four rivets *F*, *G*, *H*, and *I*, each with an area = 0.40 sq in. as shown in Fig. 7-25. When a load P = 9,600 lb is applied, determine the shear stress in each of the four rivets.

7-24. Using the riveting pattern shown in Fig. 7-25, assume that rivets *G* and *I* have areas = 0.20 sq in. and rivets *F* and *H* have rivet areas = 0.40 sq in. When P = 4,800 lb, determine the shearing stress in each of the four rivets.

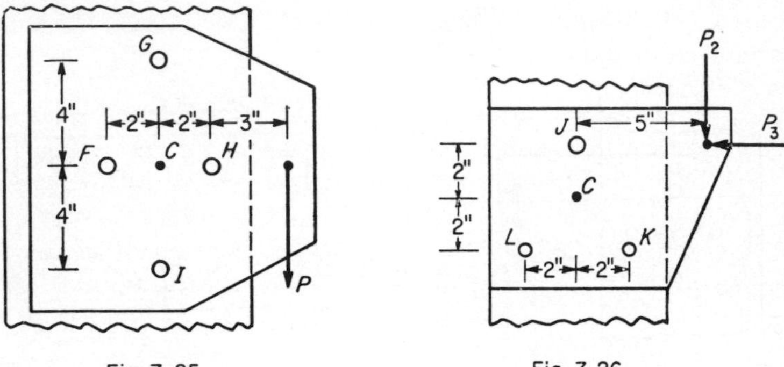

Fig. 7-25

Fig. 7-26

7-25. In Fig. 7-26, rivet *J* has an area = 0.40 sq in., and rivets *K* and *L* each have an area = 0.20 sq in. The centroid of the rivet pattern is at *C*. When P_2 = 4,800 lb and P_3 = 0, determine the shear stress in each of the rivets.

7-26. In Fig. 7-26, let P_2 = 4,800 lb and P_3 = 4,800 lb, with all other data given in Prob. 7-25 remaining the same. Determine the shearing stress on each rivet.

CHAPTER 8
COLUMNS

8-1. Columns. The configuration of the compression-loaded block which was analyzed in Chap. 1, and was also assumed to be used in Chap. 7, is shown in Fig. 8-1. While the analyses always considered elastic conditions of loading, Fig. 8-1 shows a typical compression failure for a piece of timber. The balls at the ends of the member permit the piece of material to bend laterally in any direction but prevent the lateral displacement of the top and bottom ends of the

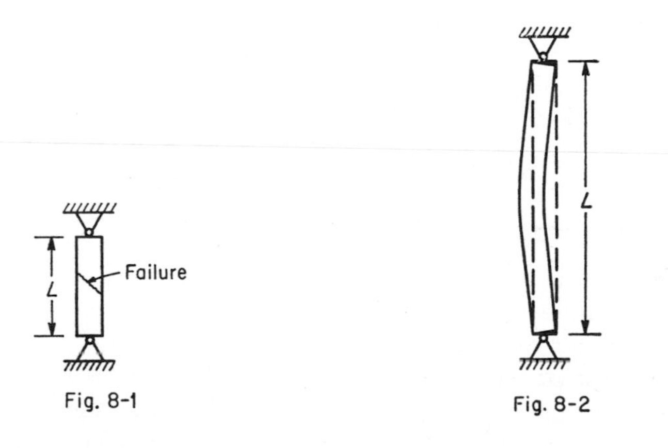

Fig. 8-1 Fig. 8-2

piece. When the ratio of the length of the block from end to end is small relative to the least cross-sectional dimension, the block remains straight, and its behavior is that of an axially loaded compression member.

When the ratio of the length of the piece of material is large compared to the least cross-sectional dimension and it is loaded by compressive forces, the behavior is similar to that shown in Fig. 8-2, and the structural piece is called a *column*. The lateral bending is called *buckling*, and the critical or buckling load occurs at the instant before the buckling action commences.

From the preceding explanation, it must be understood that the design and selection of columns cannot be made by the application of the axial-load equations.

8-2. Steel Columns. The 1963 AISC design specifications and method of column selection require considerable detailed computations. For this single reason, no attempt will be made to include this design procedure in this text.

To keep within the level and scope of this text, it will be sufficient to make use of the previous design method and equation for structural steel columns. All methods of design for all the usual column materials make use of an equation or mathematical relation which reduces the allowable axial compressive stress. The reduction factor always includes, directly or indirectly, a ratio which contains the length L and a factor determined from the cross-sectional dimensions. The allowable loads by the 1963 specifications are somewhat larger than those which were allowed by the previous design specification. This is the result of a greater theoretical study of column behavior, more uniform material, and the resulting use of higher allowable stresses.

The American Institute of Steel Construction recommends that the following formula, which has been widely adopted, be used for the design of steel columns:

$$\frac{P}{A} = 17{,}000 - 0.485\,\frac{L^2}{r^2}$$

where P = total (axial) load to be supported, lb
A = cross-sectional area, sq in.
L = length of the column, in.
r = least radius of gyration, in.

The least radius of gyration is equivalent to the square root of the least moment of inertia divided by the cross-sectional area, or

$$r = \sqrt{\frac{I}{A}}$$

In recent years, designers and steel companies have developed a large series of H steel sections, similar to I beams, but with the flange width and depth equal or nearly so. These are listed in the Appendix, along with the sections which are used entirely as beams. The sizes used as columns are

14 × 16	10 × 10
14 × 14½	8 × 8
12 × 12	

There is no reason why any of the other sections in the Appendix but not tabulated above could not be used as columns except for the difficulty in fabrication and lack of economy.

For any section, using the 14-in. × 16-in. at 246-lb-per-ft WF section as an example, radii of gyration about the X-X axis and about the Y-Y axis are listed. The columns will fail in such a way that the bending or buckling will have the Y-Y axis as the neutral axis, and as the radius of gyration for this axis is the *least* of the radii of gyration, it will *always* be used.

The term L/r is called the *slenderness ratio* of a column and is limited in practice to a value not to exceed 120. In some cases where the column is of minor importance in the structure, values of L/r may exceed 120 although good practice recommends that the limit of 120 not be exceeded.

The selection of a structural steel column is made by a method of trial.

Example. What size column is necessary to support an axial load of 150,000 lb if the length of the column is 10 ft?

This information is tabulated in the American Institute of Steel Construction Handbook. The principle of determining the proper size is a method of trial, but certain factors reduce the number of available sizes to a few.

From the equation

$$\frac{P}{A} = 17{,}000 - 0.485\frac{L^2}{r^2}$$

it will be seen that the value of P/A must be less than 17,000 and will approximate 15,000. Then, when $P = 150{,}000$ lb and P/A is approximately 15,000, A will equal $P/15{,}000$, or

$$A = \frac{150{,}000}{15{,}000} = 10 \text{ sq in. (approximate area required)}$$

Looking in the WF sections list, try the smallest column which has the approximate area. This is 8 in. × 8 in. weighing 35 lb per ft with an area of 10.30 sq in. For this size and weight, the radius of gyration r about the Y-Y axis is given as 2.03. (It should be noted that the values of r about the Y-Y axis for all the 8 in. × 8 in. weights do not vary greatly, and hence the term $0.485\ L^2/r^2$ will only change slightly for these weights.)

Then, substituting in the equation,

$$\frac{P}{A} = 17{,}000 - 0.485\frac{L^2}{r^2}$$

$$P = A\left(17{,}000 - 0.485\frac{L^2}{r^2}\right)$$

$$P = 10.30\left[17{,}000 - 0.485\frac{(120)^2}{(2.03)^2}\right]$$

$$= 10.30(17{,}000 - 1{,}690)$$

$$= 10.30(15{,}310)$$

$$= 157{,}000 \text{ lb}$$

This size WF section is safe, as it will support 157,000 lb, and the load to be supported is 150,000 lb. Also

$$\frac{L}{r} = \frac{120}{2.03} = 59.1$$

which is less than the specified maximum.

A trial should be made to see if the next lighter section which is 8 in. × 8 in. weighing 33 lb per ft will be satisfactory. Then

$$A = 9.70 \text{ sq in.}; \quad r = 2.02$$

$$P = A\left(17{,}000 - 0.485\frac{L^2}{r^2}\right)$$

$$= 9.70\left[17{,}000 - 0.485\frac{(120)^2}{(2.02)^2}\right]$$

$$= 9.70(17{,}000 - 1710)$$

$$= 9.70(15{,}290)$$

$$= 148{,}200 \text{ lb}$$

which is less than the load to be supported and hence not safe.

We know now that the 8-in. × 8-in. at 35-lb-per-ft WF column is safe and should be used. A larger size would be uneconomical because of the increased weight.

The same formula, $P/A = 17{,}000 - 0.485L^2/r^2$, is recommended for the design of steel columns having rectangular or circular cross sections. The equations for the radius of gyration for various shaped areas are listed in the Appendix under the title Elements of Sections.

In this section, it will be seen that the radius of gyration for a solid rectangle about the X-X axis is given as

$$r_{x\text{-}x} = r_{1\text{-}1} = \frac{d}{(12)^{1/2}}$$

where d = larger dimension in inches. The Y-Y axis (of the *least* radius of gyration) is now shown, but the radius of gyration about this axis is

$$r_{y\text{-}y} = r_{2\text{-}2} = \frac{b}{(12)^{1/2}}$$

where b is the least dimension in inches.

For a solid circular cross section,

$$r_{x\text{-}x} = r_{y\text{-}y} = r_{1\text{-}1} = r_{2\text{-}2} = \frac{d}{4}$$

where d = diameter in inches.

For a hollow circular cross section,

$$r_{x\text{-}x} = r_{y\text{-}y} = r_{1\text{-}1} = r_{2\text{-}2} = \frac{(d^2 + d_1{}^2)^{1/2}}{4}$$

where d is the external diameter and d_1 is the internal diameter.

Example. What axial load will a steel column $1\frac{1}{2}$ in. × 2 in. × 36 in. long safely support?

$$r_{y\text{-}y} = r_{2\text{-}2} = \frac{b}{(12)^{1/2}} = \frac{1.5}{(12)^{1/2}} - 0.433 \text{ in.}$$

To check L/r,

$$\frac{L}{r} = \frac{36}{0.433} = 83$$

which is less than 120. Then

$$\frac{P}{A} = 17{,}000 - 0.485\frac{L^2}{r^2}$$

$$P = A\left(17{,}000 - 0.485\frac{L^2}{r^2}\right)$$

which is equivalent to writing the formula

$$P = A\left[17{,}000 - 0.485\left(\frac{L}{r}\right)^2\right]$$

$$= 2 \times 1.5[17{,}000 - 0.485(83)^2]$$

$$= 3(17{,}000 - 3340)$$

$$= 3 \times 13{,}660$$

$$= 40{,}980 \text{ lb safe load}$$

Example. A hollow-pipe column is 10 ft long. The external diameter is 4.50 in. and the internal diameter is 4.026 in. (These are the diameters of 4-in. standard pipe.) What safe concentric (axial) load will the column support?

$$r_{y\text{-}y} = r_{2\text{-}2} = \frac{(d^2 + d_1^2)^{1/2}}{4} = \frac{[(4.50)^2 + (4.026)^2]^{1/2}}{4}$$

$$r_{2\text{-}2} = \frac{(36.46)^{1/2}}{4} = 1.51 \text{ in.}$$

$$A = \frac{\pi}{4}d^2 - \frac{\pi}{4}d_1^2 = \frac{\pi}{4}(d^2 - d_1^2) = \frac{\pi}{4}(20.25 - 16.21)$$

$$= \frac{\pi}{4}(4.04) = 3.17 \text{ sq in.}$$

To check $\frac{L}{r}$,

$$\frac{L}{r} = \frac{120 \text{ in.}}{1.51 \text{ in.}} = 79.5$$

which is less than 120. Then

$$P = A\left(17{,}000 - 0.485\frac{L^2}{r^2}\right)$$

$$= A\left[17{,}000 - 0.485\left(\frac{L}{r}\right)^2\right]$$

$$= 3.17[17{,}000 - 0.485(79.5)^2]$$

$$= 3.17(17{,}000 - 3{,}060)$$

$$= 3.17 \times 13{,}940$$

$$= 44{,}100 \text{ lb safe load}$$

In the design of machine parts, very often other equations are specified and must be used. Such a design entails no additional difficulties as the same general procedure is followed, except that a value of L/r greater than 120 may be permitted.

8-3. Structural Aluminum Columns. The aluminum industry, in conjunction with the cooperating national engineering societies, has recommended a design equation for structural aluminum columns. This equation is in the form of

$$\frac{P}{A} = \frac{E\pi^2}{(KL/r)^2}$$

and with the recommended factor of safety of 2.5 becomes

$$\frac{P}{A} = \frac{E\pi^2}{2.5(KL/r)^2}$$

where K is a constant which reflects the end restraint conditions, and r is the least radius of gyration. If the columns are used where there is no end restraint as to the direction of buckling, $K = 1$. However, it is difficult to obtain the idealized condition of no restraint, and the *recommended value* of $K = 0.75$. The preceding equation is recommended for values of KL/r less than 90 and greater than 63. The minimum L/r value occurs when the P/A ratio equals 63 for the aluminum alloy 6061 T6. This alloy of aluminum is one of the structural alloys.

When the KL/r ratio is less than 63, and the factor of safety is 2.5, the design equation is

$$\frac{2.5P}{A} = 38{,}300 - 202\,\frac{KL}{r}$$

Values of the constants for other aluminum alloys are readily available in handbooks.

All design theory and recommended equations with their limitations are the result of considerable theoretical study and practical full-size structural tests. As a terminal result, the design equations which are included in this text represent the recommended design practice. Considerable judgment on the part of the designer is necessary, as he must decide which theory to use and what values of the constants reflect the end conditions and material.

8-4. Timber Columns. The design and selection of a timber column require a slightly different procedure than does the design of a steel column. The most recent and thorough study of timber columns has been made by the Forest Products Laboratory of the United States Forest Service. The results of this study have been adopted by the American Society of Testing Materials and are recommended as good practice for general use.

Timber columns are classified according to the ratio L/d, where L is the length in inches and d is the least dimension of the cross section in inches.

Class 1. When the value of L/d does not exceed 10, the timber post (column) shall be designed using the compression equation

$$S = \frac{P}{A}$$

where S = safe working stress of the timber, lb per sq in. (see Table 8-1)
P = total load, lb
A = cross-sectional area, sq in.

TABLE 8-1. SAFE WORKING STRESSES, LB PER SQ IN.
Common Structural Grade, Continuously Dry

Species	Values of L/d											
	10 or less	12	14	16	18	20	25	30	35	40	50	E*
Douglas fir:												
Coast region	880	870	861	847	826	796	675	487	358	274	175	1.6
Dense	1025	1017	996	965	935	893	698	487	358	274	175	1.6
Hemlock:												
West coast	720	712	706	696	680	660	573	426	313	240	153	1.4
Larch:												
Western	880	863	849	828	798	752	570	396	291	223	142	1.3
Pine:												
Southern	880	870	861	847	826	796	675	487	358	274	175	1.6
Dense	1025	1017	996	965	935	893	698	487	358	274	175	1.6
Spruce:												
White	640	632	627	617	602	582	500	365	268	206	132	1.2
Oak:												
Red or white	800	791	783	771	753	729	626	457	336	257	164	1.5

* Values of E are to be multiplied by 1,000,000. Thus for west coast hemlock,

$$E = 1.4 \times 1{,}000{,}000 = 1{,}400{,}000.$$

The student is referred to a handbook for similar tables of safe working stresses for other grade timber columns which are to be used where the conditions are occasionally wet or continuously wet.

Example of Class 1 Timber Column. A common-grade pine timber post is 3 ft long and 4 in. × 4 in. in cross section. What safe load can it support?

$$\frac{L}{d} = \frac{36 \text{ in.}}{4 \text{ in.}} = 9$$

Therefore, this is a class 1 column.

From Table 8-1, opposite southern pine, find 880 psi in the column of figures for L/d of 10 or less.

Then

$$S = \frac{P}{A} \qquad \text{and} \qquad P = SA$$

Hence

$$P = 880 \times 4 \times 4 = 14{,}080 \text{ lb}$$

which is the safe working load that the post will support if the common-grade southern pine is continuously dry.

Class 2. When the value of L/d is greater than 10 but less than 25 (intermediate length columns), the following formula shall apply:

$$\frac{P}{A} = S\left[1 - \frac{1}{3}\left(\frac{L}{kd}\right)^4\right]$$

where P = total load, lb

A = cross-sectional area, sq in.

S = safe working stress for short (L/d less than 10) columns (see note following)

L = length, in.

d = least dimension of cross section, in.

k = a constant, depending on the grade and kind of timber and the conditions under which the column is to be used (see Table 8-2)

The formula above is called the Forest Products Laboratory formula for intermediate length columns; this is abbreviated FPL intermediate column formula.

Note. The meaning of S will appear inconsistent with the class of columns to which this formula applies, but S correctly represents the safe working stress for a short column. Inasmuch as the numerical result of the term within the brackets is always less than one, this result is the factor by means of which the safe working stress for a short column is reduced so that it correctly applies to an intermediate length column. These facts are illustrated in the following numerical example by two solutions for the same given data.

TABLE 8-2. VALUES OF k FOR TIMBER COLUMNS OF INTERMEDIATE LENGTH
Common Structural Grade, Continuously Dry

Species	k
Douglas fir:	
Coast region	27.3
Dense	25.3
Hemlock:	
West coast	28.3
Larch:	
Western	24.6
Pine:	
Southern	27.3
Dense	25.3
Spruce:	
White	27.8
Oak:	
Red or white	27.8

The student is referred to a handbook for values of k for other grades of all species and for other conditions of use. However, the values of safe working stresses, as given in Table 8-1, will eliminate the major computations required when using the FPL intermediate column formula. Thus, the formula becomes

$$\frac{P}{A} = S$$

where S is the safe working stress from Table 8-1 for the proper species and value of L/d.

Example of Class 2 Timber Column. A common-grade southern pine column is 6 in. × 8 in. × 10 ft long. What load will it safely support?

$$\frac{L}{d} = \frac{10 \times 12}{6} = \frac{120}{6} = 20$$

Therefore, this is a class 2 column.

Using $$\frac{P}{A} = S$$

From Table 8-1, opposite southern pine, find 796 psi in the column of figures for $L/d = 20$. Then

$$P = AS = 6 \times 8 \times 796 = 38{,}208 \text{ lb (safe load)}$$

To check, and show the use of the complete FPL formula for intermediate columns,

$$\frac{P}{A} = S\left[1 - \frac{1}{3}\left(\frac{L}{kd}\right)^4\right]$$

From Table 8-1

$$S = 880 \text{ psi for short columns}$$

From Table 8-2

$$k = 27.3$$

$$A = 6 \times 8 = 48 \text{ sq in.}$$

$$\frac{L}{d} = 20$$

Therefore

$$P = AS\left[1 - \frac{1}{3}\left(\frac{L}{kd}\right)^4\right]$$

$$= 48 \times 880\left[1 - \frac{1}{3}\left(\frac{20}{27.3}\right)^4\right]$$

$$= 48 \times 880[1 - \tfrac{1}{3}(0.288)]$$

$$= 42{,}240(1 - 0.096)$$

$$= 42{,}240 \times 0.904 = 38{,}185 \text{ lb (which checks closely enough)}$$

The correct values of S for values of L/d not given in the table may be found by interpolation as shown in the following example.

Example. What load can a common-grade southern pine column 6 in. × 8 in. × 11 ft long safely support?

$$\frac{L}{d} = \frac{11 \times 12}{6} = 22$$

In Table 8-1, for southern pine, when

$$\frac{L}{d} = 20 \qquad S = 796 \text{ psi}$$

and when

$$\frac{L}{d} = 25 \qquad S = 675 \text{ psi}$$

It will be seen that 22 is two-fifths of the way between 20 and 25. Therefore two-fifths of the difference in the values of $S = 796$ and $S = 675$ is $\frac{2}{5} \times 121 = \frac{242}{5} = 48.4$, which is to be subtracted from the value of $S = 796$ when $L/d = 20$. Hence

$$S = 796 - 48.4 = 747.6 \text{ psi} \qquad \text{when} \qquad \frac{L}{d} = 22$$

Then, using

$$\frac{P}{A} = S$$

$$\begin{aligned} P &= AS \\ &= 6 \times 8 \times 747.6 \\ &= 35{,}885 \text{ lb (safe load)} \end{aligned}$$

Class 3. When the value of L/d is 25 or greater, but less than 50 (long columns), the following formula shall apply:

$$\frac{P}{A} = \frac{E\pi^2}{36(L/d)^2}$$

where P = total load, lb
A = cross-sectional area, sq in.
E = modulus of elasticity (Table 8-1)
L = length, in.
d = least dimension, in.

This formula is called the Forest Products Laboratory formula for long columns. It is abbreviated FPL long column formula.

Note. Timber columns shall be limited to a maximum value of $L/d = 50$. The student is referred to a handbook for values of E for other species of timber and for other conditions of use. However, Table 8-1 will again eliminate the computations necessary with this (FPL long column) formula by giving correct, safe working stresses when L/d exceeds 25. Thus

$$\frac{P}{A} = S$$

where S is the safe working stress from Table 8-1 for the proper species and value of L/d.

Example of Class 3 Timber Column. A common-grade southern pine column is 6 in. × 8 in. × 15 ft long. What load will it safely support?

$$\frac{L}{d} = \frac{15 \times 12}{6} = 30 \text{ (therefore, this is a class 3 column)}$$

Using

$$\frac{P}{A} = S$$

From Table 8-1, opposite southern pine, find 487 psi in the column of figures for $L/d = 30$.

Then

$$P = AS = 6 \times 8 \times 487 = 23{,}380 \text{ lb (safe load)}$$

To check, and show the use of the complete FPL long column formula

$$\frac{P}{A} = \frac{E\pi^2}{36(L/d)^2}$$

From Table 8-1, for southern pine, $E = 1{,}600{,}000$ psi.

$$A = 48 \text{ sq in.}$$

$$\frac{L}{d} = 30$$

Therefore

$$P = \frac{A\pi^2 E}{36(L/d)^2} = \frac{48\pi^2 \times 1{,}600{,}000}{36 \times (30)^2}$$

$$= 23{,}380 \text{ lb (safe load)}$$

Problems

8-1. A cylindrical steel rod ⅜ in. in diameter and 24 in. long is to be used as a compression member. Assuming that it acts as a column, what safe load would you recommend?

8-2. If the length of the bar in Prob. 8-1 is doubled, what maximum load will it support as a compression member?

8-3. A solid steel bar 1 in. × 2 in. is used as a compression member which has a length of 15 in. What is the safe load?

8-4. A 4-in. extra-strong steel pipe has diameters of 4.500 in. and 3.826 in. with an area of 4.41 sq in. $I = 9.61$ in.^{4} If L/r is limited to 100, what is the maximum length of pipe which can be used? What is the safe load capacity of the pipe when it is used as a compression member of maximum length?

8-5. A 6-in. × 6-in. square structural steel tube with a wall thickness of 0.50 in. has an area = 10.14 sq in. and a radius of gyration $r = 2.18$ in. What is the maximum safe load which the tube can support as a column if it is 15 ft long?

8-6. Design the steel compression member *DE* in Fig. 1-17. The only specification limits the size to a WF 8-in. by 8-in. section. Assume that the AISC formula is applicable.

8-7. What 8-in. WF steel size would you recommend for member *EF* in the truss shown in Fig. 1-17?

8-8. In Fig. 8-3, the dimensions are $BF = 8$ ft; $AB = BC = 6$ ft; $CD = CE = 10$ ft. The tripod is used to support the 400-kip load applied vertically at *F*. Determine the required 8-in. × 8-in. WF steel section for member *AF*.

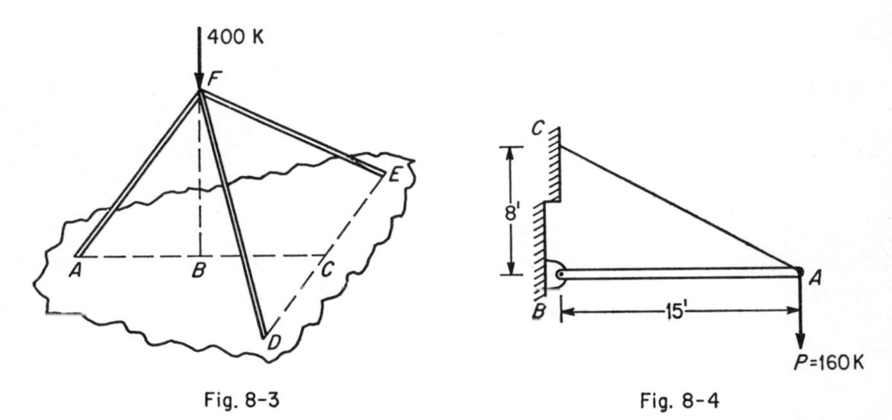

Fig. 8-3 Fig. 8-4

8-9. In Fig. 8-3, determine the required 8-in. × 8-in. WF steel section for members *DF* and *EF*. Refer to Prob. 8-8 for the dimensions of the tripod.

8-10. When $P = 160$ kips is applied at *A* in Fig. 8-4, determine the required 10-in. × 10-in. WF steel section for member *AB*.

8-11. A steel interior building column is 12 ft long and supports a floor area 20 ft × 25 ft. The design load on the floor is 550 lb per sq ft, and the floor system is assumed to weigh 50 lb per sq ft. Determine the required 8-in. × 8-in. or 10-in. × 10-in. WF section to support the floor area safely.

Problems 8-12 to 8-18 refer to aluminum-alloy columns. For the alloy, $B = 38{,}300$ psi, $D = 202$, and $C = 63$. The factor of safety is 2.5, and $K = \frac{3}{4}$. Use $E\pi^2 = 10.2(10)^7$ psi.

8-12. A 6-in. × 6-in. extruded aluminum-alloy column has $r_{y\text{-}y} =$ 1.45 in. and $A = 5.40$ sq in. Determine the safe compressive column load when $L = 12$ ft.

8-13. If the length of the column is decreased to 6 ft in Prob. 8-12, determine the safe load.

8-14. There is a choice of three 8-in. × 8-in. aluminum-alloy columns. The necessary data for each are:

1. $A = 10.72$ sq in.; $r_{y\text{-}y} = 2.01$ in.
2. $A = 11.24$ sq in.; $r_{y\text{-}y} = 1.88$ in.
3. $A = 12.99$ sq in.; $r_{y\text{-}y} = 1.82$ in.

Which of the three available sizes would you select to support a compressive column load $P = 65$ kips on a length of 18 ft?

8-15. Two aluminum-alloy I beams are to be used as a column, as shown in Fig. 8-5. They are held together by bars which are not shown to form a unit structural column. The data for *each* I section are $A = 8.8$ sq in.; $I_x = 134$ in.^{4}; $I_y' = 7.5$ in.^{4} Determine the distance d so that the total I_x is equal to the total I_y, and $r_{x\text{-}x}$ equals $r_{y\text{-}y}$. This implies that buckling would start about either the X-X or Y-Y axis.

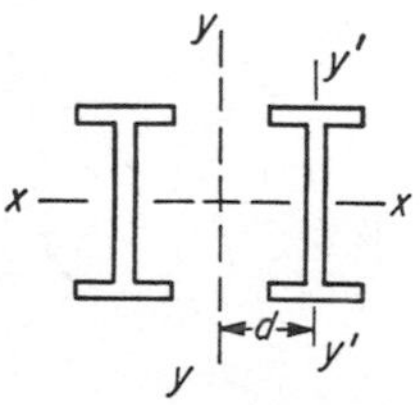

Fig. 8-5

8-16. When the column, for which the pertinent data are given in Prob. 8-15, has a length of 15 ft, determine the maximum permissible concentric load.

8-17. When the column length is 20 ft, determine the allowable concentric load. All other data are the same as those given in Prob. 8-15.

8-18. Determine the limiting value of KL/r for an aluminum-alloy column when the value of $P/A = 20{,}000$ psi. Use a factor of safety of 1.

The moisture condition of uses of the timber columns referred to in the following problems is assumed to be continuously dry. The timber is classed as common structural grade. All columns or posts are full size.

8-19. A hemlock post is 2 in. × 4 in. × 4 ft long. Determine the maximum safe end-compressive load. Solve by the proper equation, and check by using the safe stress from Table 8-1.

8-20. If the length of the post in Prob. 8-19 is increased to 8 ft, determine the safe load. Solve by the proper equation, and check by using the safe stress from Table 8-1.

8-21. If the length of the post in Prob. 8-19 is decreased in length to 2 ft, determine the safe load.

8-22. Hemlock with a *square* cross section is to be used for the compression members of the simple truss. It should be assumed that the compression members act as columns, with member AC being one single piece of timber 12 ft long. See Fig. 8-6.

Based on the *long-column equation* for the design, determine the correct size for member AB. Is the assumption that the long column equation applies to the design correct?

8-23. Using the dimension of the column calculated in Prob. 8-22 and increasing it to the next larger full inch (Example: If $d = 2.73$ in.,

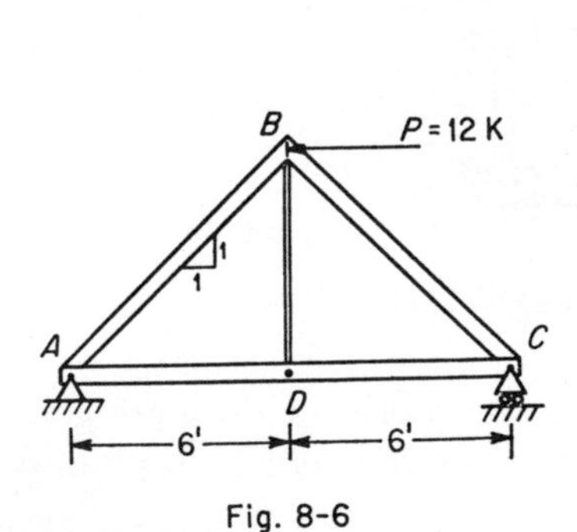

Fig. 8-6

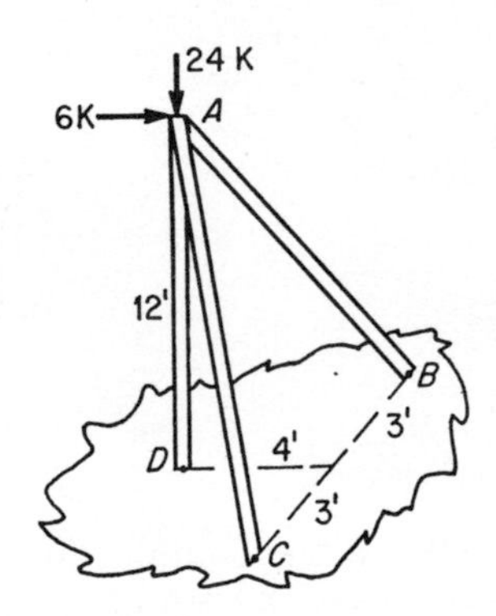

Fig. 8-7

use $d = 3$ in.), determine the *greater* dimension of the compression member AC in Fig. 8-6. (Assume a long-column design condition.)

8-24. A tripod stand is used to support the two loads shown in Fig. 8-7. Assume that the compression members act as concentrically loaded columns. Using the long-column equation, determine the correct size of a square piece of southern pine for member AD.

8-25. In Fig. 8-7, determine the size of square southern pine timbers for members AB and AC if they are compression members. Use the long-column equation for the design.

If the members are in tension, determine the size of square southern pine timbers using a safe tensile stress of 880 psi.

APPENDIX

Properties of WF Sections

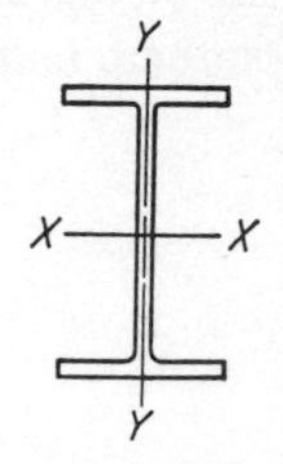

Nominal size, in.	Weight per foot, lb	Area, in.²	Depth, in.	Flange		Web thickness, in.	Axis Y-X			Axis Y-Y		
				Width, in.	Thickness, in.		I, in.⁴	S, in.³	r, in.	I, in.⁴	S, in.³	r, in.
36 × 16½	300	88.17	36.72	16.655	1.680	0.945	20290.2	1105.1	15.17	1225.2	147.1	3.73
	280	82.32	36.50	16.595	1.570	0.885	18819.3	1031.2	15.12	1127.5	135.9	3.70
	260	76.56	36.24	16.555	1.440	0.845	17233.8	951.1	15.00	1020.6	123.3	3.65
	245	72.03	36.06	16.512	1.350	0.802	16092.2	892.5	14.95	944.7	114.4	3.62
	230	67.73	35.88	16.475	1.260	0.765	14988.4	835.5	14.88	870.9	105.7	3.59
36 × 12	194	57.11	36.48	12.117	1.260	0.770	12103.4	663.6	14.56	355.4	58.7	2.49
	182	53.54	36.32	12.072	1.180	0.725	11281.5	621.2	14.52	327.7	54.3	2.47
	170	49.98	36.16	12.027	1.100	0.680	10470.0	579.1	14.47	300.6	50.0	2.45
	160	47.09	36.00	12.000	1.020	0.653	9738.8	541.0	14.38	275.4	45.9	2.42
	150	44.16	35.84	11.972	0.940	0.625	9012.1	502.9	14.29	250.4	41.8	2.38
	135	39.70	35.55	11.945	0.794	0.598	7796.1	438.6	14.01	207.1	34.7	2.28
33 × 15¾	240	70.52	33.50	15.865	1.400	0.830	13585.1	811.1	13.88	874.3	110.2	3.52
	220	64.73	33.25	15.810	1.275	0.775	12312.1	740.6	13.79	782.4	99.0	3.48
	200	58.79	33.00	15.750	1.150	0.715	11048.2	669.6	13.71	691.7	87.8	3.43
33 × 11½	152	44.71	33.50	11.565	1.055	0.635	8147.6	486.4	13.50	256.1	44.3	2.39
	141	41.51	33.31	11.535	0.960	0.605	7442.2	446.8	13.39	229.7	39.8	2.35
	130	38.26	33.10	11.510	0.855	0.580	6699.0	404.8	13.23	201.4	35.0	2.29
	118	34.71	32.86	11.484	0.738	0.554	5886.9	358.3	13.02	170.3	29.7	2.22
30 × 15	210	61.78	30.38	15.105	1.315	0.775	9872.4	649.9	12.64	707.9	93.7	3.38
	190	55.90	30.12	15.040	1.185	0.710	8825.9	586.1	12.57	624.6	83.1	3.34
	172	50.65	29.88	14.985	1.065	0.655	7891.5	528.2	12.48	550.1	73.4	3.30

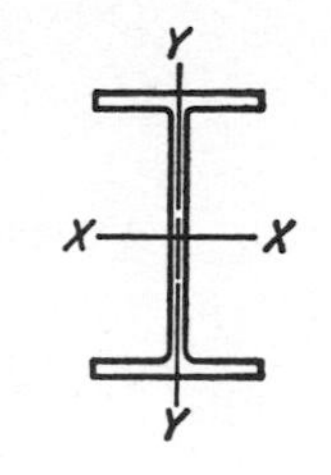

PROPERTIES OF WF SECTIONS (*Continued*)

Nominal size, in.	Weight per foot, lb	Area, in.²	Depth, in.	Flange		Web thickness, in.	Axis X-X			Axis Y-Y		
				Width, in.	Thickness, in.		I, in.⁴	S, in.³	r, in.	I, in.⁴	S, in.³	r, in.
30 × 10½	132	38.83	30.30	10.551	1.000	0.615	5753.1	379.7	12.17	185.0	35.1	2.18
	124	36.45	30.16	10.521	0.930	0.585	5347.1	354.6	12.11	160.7	[illegible]	2.10
	116	34.13	30.00	10.500	0.850	0.564	4919.1	327.9	12.00	153.2	29.2	2.12
	108	31.77	29.82	10.484	0.760	0.548	4461.0	299.2	11.85	135.1	25.8	2.06
	99	29.11	29.64	10.458	0.670	0.522	3988.6	269.1	11.70	116.9	22.4	2.00
27 × 14	177	52.10	27.31	14.090	1.190	0.725	6728.6	492.8	11.36	518.9	73.7	3.16
	160	47.04	27.08	14.023	1.075	0.658	6018.6	444.5	11.31	458.0	65.3	3.12
	145	42.68	26.88	13.965	0.975	0.600	5414.3	402.9	11.26	406.9	58.3	3.09
27 × 10	114	33.53	27.28	10.070	0.932	0.570	4080.5	299.2	11.03	149.6	29.7	2.11
	102	30.01	27.07	10.018	0.827	0.518	3604.1	266.3	10.96	129.5	25.9	2.08
	94	27.65	26.91	9.990	0.747	0.490	3266.7	242.8	10.87	115.1	23.0	2.04
	84	24.71	26.69	9.963	0.636	0.463	2824.8	211.7	10.69	95.7	19.2	1.97
24 × 14	160	47.04	24.72	14.091	1.135	0.656	5110.3	413.5	10.42	492.6	69.9	3.23
	145	42.62	24.49	14.043	1.020	0.608	4561.0	372.5	10.34	434.3	61.8	3.19
	130	38.21	24.25	14.000	0.900	0.565	4009.5	330.7	10.24	375.2	53.6	3.13
24 × 12	120	35.29	24.31	12.088	0.930	0.556	3635.3	299.1	10.15	254.0	42.0	2.68
	110	32.36	24.16	12.042	0.855	0.510	3315.0	274.4	10.12	229.1	38.0	2.66
	100	29.43	24.00	12.000	0.775	0.468	2987.3	248.9	10.08	203.5	33.9	2.63
24 × 9	94	27.63	24.29	9.061	0.872	0.516	2683.0	220.9	9.85	102.2	22.6	1.92
	84	24.71	24.09	9.015	0.772	0.470	2364.3	196.3	9.78	88.3	19.6	1.89
	76	22.37	23.91	8.985	0.682	0.440	2096.4	175.4	9.68	76.5	17.0	1.85
	68	20.00	23.71	8.961	0.582	0.416	1814.5	153.1	9.53	63.8	14.2	1.79
21 × 13	142	41.76	21.46	13.132	1.095	0.659	3403.1	317.2	9.03	385.9	58.8	3.04
	127	37.34	21.24	13.061	0.985	0.588	3017.2	284.1	8.99	338.6	51.8	3.01
	112	32.93	21.00	13.000	0.865	0.527	2620.6	249.6	8.92	289.7	44.6	2.96
21 × 9	96	28.21	21.14	9.038	0.935	0.575	2088.9	197.6	8.60	109.3	24.2	1.97
	82	24.10	20.86	8.962	0.795	0.499	1752.4	168.0	8.53	89.6	20.0	1.93
21 × 8¼	73	21.46	21.24	8.295	0.740	0.455	1600.3	150.7	8.64	66.2	16.0	1.76
	68	20.02	21.13	8.270	0.685	0.430	1478.3	139.9	8.59	60.4	14.6	1.74
	62	18.23	20.99	8.240	0.615	0.400	1326.8	126.4	8.53	53.1	12.9	1.71
	55	16.18	20.80	8.215	0.522	0.375	1140.7	109.7	8.40	44.0	10.7	1.65

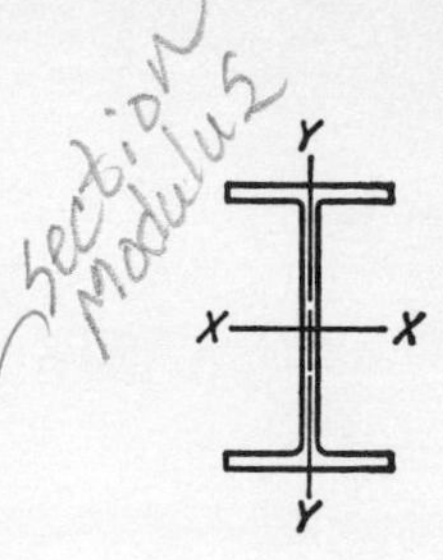

PROPERTIES OF WF SECTIONS (*Continued*)

Nominal size, in.	Weight per foot, lb	Area, in.^{2}	Depth, in.	Flange		Web thickness, in.	Axis X-X			Axis Y-Y		
				Width, in.	Thickness, in.		I, in.^{4}	S, in.^{3}	r, in.	I, in.^{4}	S, in.^{3}	r, in.
18 × 11¾	114	33.51	18.48	11.833	0.991	0.595	2033.8	220.1	7.79	255.6	43.2	2.76
	105	30.86	18.32	11.792	0.911	0.554	1852.5	202.2	7.75	231.0	39.2	2.73
	96	28.22	18.16	11.750	0.831	0.512	1674.7	184.4	7.70	206.8	35.2	2.71
18 × 18¾	85	24.97	18.32	8.838	0.911	0.526	1429.9	156.1	7.57	99.4	22.5	2.00
	77	22.63	18.16	8.787	0.831	0.475	1286.8	141.7	7.54	88.6	20.2	1.98
	70	20.56	18.00	8.750	0.751	0.438	1153.9	128.2	7.49	78.5	17.9	1.95
	64	18.80	17.87	8.715	0.686	0.403	1045.8	117.0	7.46	70.3	16.1	1.93
18 × 7½	60	17.64	18.25	7.558	0.695	0.416	984.0	107.8	7.47	47.1	12.5	1.63
	55	16.19	18.12	7.532	0.630	0.390	889.9	98.2	7.41	42.0	11.1	1.61
	50	14.71	18.00	7.500	0.570	0.358	800.6	89.0	7.38	37.2	9.9	1.59
	45	13.24	17.86	7.477	0.499	0.335	704.5	78.9	7.30	31.9	8.5	1.55
16 × 11½	96	28.22	16.32	11.533	0.875	0.535	1355.1	166.1	6.93	207.2	35.9	2.71
	88	25.87	16.16	11.502	0.795	0.504	1222.6	151.3	6.87	185.2	32.2	2.67
16 × 8½	78	22.92	16.32	8.586	0.875	0.529	1042.6	127.8	6.74	87.5	20.4	1.95
	71	20.86	16.16	8.543	0.795	0.486	936.9	115.9	6.70	77.9	18.2	1.93
	64	18.80	16.00	8.500	0.715	0.443	833.8	104.2	6.66	68.4	16.1	1.91
	58	17.04	15.86	8.464	0.645	0.407	746.4	94.1	6.62	60.5	14.3	1.88
16 × 7	50	14.70	16.25	7.073	0.628	0.380	655.4	80.7	6.68	34.8	9.8	1.54
	45	13.24	16.12	7.039	0.563	0.346	583.3	72.4	6.64	30.5	8.7	1.52
	40	11.77	16.00	7.000	0.503	0.307	515.5	64.4	6.62	26.5	7.6	1.50
	36	10.59	15.85	6.992	0.428	0.299	446.3	56.3	6.49	22.1	6.3	1.45
14 × 16	426	125.25	18.69	16.695	3.033	1.875	6610.3	707.4	7.26	2359.5	282.7	4.34
	398	116.98	18.31	16.590	2.843	1.770	6013.7	656.9	7.17	2169.7	261.6	4.31
	370	108.78	17.94	16.475	2.658	1.655	5454.2	608.1	7.08	1986.0	241.1	4.27
	342	100.59	17.56	16.365	2.468	1.545	4911.5	559.4	6.99	1806.9	220.8	4.24
	314	92.30	17.19	16.235	2.283	1.415	4399.4	511.9	6.90	1631.4	201.0	4.20
	287	84.37	16.81	16.130	2.093	1.310	3912.1	465.5	6.81	1466.5	181.8	4.17
	264	77.63	16.50	16.025	1.938	1.205	3526.0	427.4	6.74	1331.2	166.1	4.14
	246	72.33	16.25	15.945	1.813	1.125	3228.9	397.4	6.68	1226.6	153.9	4.12
	237	69.69	16.12	15.910	1.748	1.090	3080.9	382.2	6.65	1174.8	147.7	4.11
	228	67.06	16.00	15.865	1.688	1.045	2942.4	367.8	6.62	1124.8	141.8	4.10
	219	64.36	15.87	15.825	1.623	1.005	2798.2	352.6	6.59	1073.2	135.6	4.08
	211	62.07	15.75	15.800	1.563	0.980	2671.4	339.2	6.56	1028.6	130.2	4.07
	202	59.39	15.63	15.750	1.503	0.930	2538.8	324.9	6.54	979.7	124.4	4.06
	193	56.73	15.50	15.710	1.438	0.890	2402.4	310.0	6.51	930.1	118.4	4.05

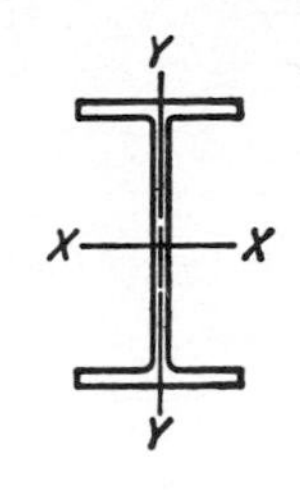

Properties of WF Sections (*Continued*)

Nominal size, in.	Weight per foot, lb	Area, in.^{2}	Depth, in.	Flange		Web thickness, in.	Axis X-X			Axis Y-Y		
				Width, in.	Thickness, in.		I, in.^{4}	S, in.^{3}	r, in.	I, in.^{4}	S, in.^{3}	r, in.
14 × 16	184	54.07	15.38	15.660	1.378	0.840	2274.8	295.8	6.49	882.7	112.7	4.04
	176	51.73	15.25	15.640	1.313	0.820	2149.6	281.9	6.45	837.9	107.1	4.02
	167	49.09	15.12	15.600	1.248	0.700	2020.8	267.3	6.42	790.2	101.3	4.01
	158	46.47	15.00	15.550	1.188	0.730	1900.6	253.4	6.40	745.0	95.8	4.00
	150	44.08	14.88	15.515	1.128	0.695	1786.9	240.2	6.37	702.5	90.6	3.99
	142	41.85	14.75	15.500	1.063	0.680	1672.2	226.7	6.32	660.1	85.2	3.97
	*320	94.12	16.81	16.710	2.093	1.890	4141.7	492.8	6.63	1635.1	195.7	4.17
14 × 14½	136	39.98	14.75	14.740	1.063	0.660	1593.0	216.0	6.31	567.7	77.0	3.77
	127	37.33	14.62	14.690	0.998	0.610	1476.7	202.0	6.29	527.6	71.8	3.76
	119	34.99	14.50	14.650	0.938	0.570	1373.1	189.4	6.26	491.8	67.1	3.75
	111	32.65	14.37	14.620	0.873	0.540	1266.5	176.3	6.23	454.9	62.2	3.73
	103	30.26	14.25	14.575	0.813	0.495	1165.8	163.6	6.21	419.7	57.6	3.72
	95	27.94	14.12	14.545	0.748	0.465	1063.5	150.6	6.17	383.7	52.8	3.71
	87	25.56	14.00	14.500	0.688	0.420	966.9	138.1	6.15	349.7	48.2	3.70
14 × 12	84	24.71	14.18	12.023	0.778	0.451	928.4	130.9	6.13	225.5	37.5	3.02
	78	22.94	14.06	12.000	0.718	0.428	851.2	121.1	6.09	206.9	34.5	3.00
14 × 10	74	21.76	14.19	10.072	0.783	0.450	796.8	112.3	6.05	133.5	26.5	2.48
	68	20.00	14.06	10.040	0.718	0.418	724.1	103.0	6.02	121.2	24.1	2.46
	61	17.94	13.91	10.000	0.643	0.378	641.5	92.2	5.98	107.3	21.5	2.45
14 × 8	53	15.59	13.94	8.062	0.658	0.370	542.1	77.8	5.90	57.5	14.3	1.92
	48	14.11	13.81	8.031	0.593	0.339	484.9	70.2	5.86	51.3	12.8	1.91
	43	12.65	13.68	8.000	0.528	0.308	429.0	62.7	5.82	45.1	11.3	1.89
14 × 6¾	38	11.17	14.12	6.776	0.513	0.313	385.3	54.6	5.87	24.6	7.3	1.49
	34	10.00	14.00	6.750	0.453	0.287	339.2	48.5	5.83	21.3	6.3	1.46
	30	8.81	13.86	6.733	0.383	0.270	289.6	41.8	5.73	17.5	5.2	1.41
12 × 12	190	55.86	14.38	12.670	1.733	1.060	1892.5	263.2	5.82	589.7	93.1	3.25
	161	47.38	13.88	12.515	1.486	0.905	1541.8	222.2	5.70	486.2	77.7	3.20
	133	39.11	13.38	12.365	1.236	0.755	1221.2	182.5	5.59	389.9	63.1	3.16
	120	35.31	13.12	12.320	1.106	0.710	1071.7	163.4	5.51	345.1	56.0	3.13
	106	31.19	12.88	12.230	0.986	0.620	930.7	144.5	5.46	300.9	49.2	3.11
	99	29.09	12.75	12.190	0.921	0.580	858.5	134.7	5.43	278.2	45.7	3.09
	92	27.06	12.62	12.155	0.856	0.545	788.9	125.0	5.40	256.4	42.2	3.08
	85	24.98	12.50	12.105	0.796	0.495	723.3	115.7	5.38	235.5	38.9	30.7
	79	23.22	12.38	12.080	0.736	0.470	663.0	107.1	5.34	216.4	35.8	3.05
	72	21.16	12.25	12.040	0.671	0.430	597.4	97.5	5.31	195.3	32.4	3.04
	65	19.11	12.12	12.000	0.606	0.390	533.4	88.0	5.28	174.6	29.1	3.02

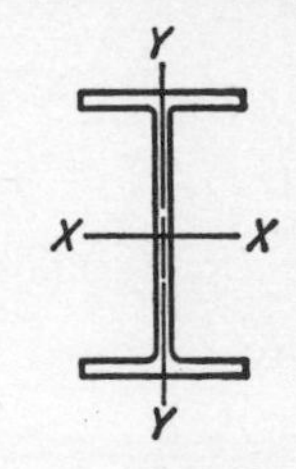

PROPERTIES OF WF SECTIONS (*Continued*)

Nominal size, in.	Weight per foot, lb	Area, in.^{2}	Depth, in.	Flange		Web thickness, in.	Axis X-X			Axis Y-Y		
				Width, in.	Thickness, in.		I, in.^{4}	S, in.^{3}	r, in.	I, in.^{4}	S, in.^{3}	r, in.
12 × 10	58	17.06	12.19	10.014	0.641	0.359	476.1	78.1	5.28	107.4	21.4	2.51
	53	15.59	12.06	10.000	0.576	0.345	426.2	70.7	5.23	96.1	19.2	2.48
12 × 8	50	14.71	12.19	8.077	0.641	0.371	394.5	64.7	5.18	56.4	14.0	1.96
	45	13.24	12.06	8.042	0.576	0.336	350.8	58.2	5.15	50.0	12.4	1.94
	40	11.77	11.94	8.000	0.516	0.294	310.1	51.9	5.13	44.1	11.0	1.94
12 × 6½	36	10.59	12.24	6.565	0.540	0.305	280.8	45.9	5.15	23.7	7.2	1.50
	31	9.12	12.09	6.525	0.465	0.265	238.4	39.4	5.11	19.8	6.1	1.47
	27	7.97	11.96	6.500	0.400	0.240	204.1	34.1	5.06	16.6	5.1	1.44
10 × 10	112	32.92	11.38	10.415	1.248	0.755	718.7	126.3	4.67	235.4	45.2	2.67
	100	29.43	11.12	10.345	1.118	0.685	625.0	112.4	4.61	206.6	39.9	2.65
	89	26.19	10.88	10.275	0.998	0.615	542.4	99.7	4.55	180.6	35.2	2.63
	77	22.67	10.62	10.195	0.868	0.535	457.2	86.1	4.49	153.4	30.1	2.60
	72	21.18	10.50	10.170	0.808	0.510	420.7	80.1	4.46	141.8	27.9	2.59
	66	19.41	10.38	10.117	0.748	0.457	382.5	73.7	4.44	129.2	25.5	2.58
	60	17.66	10.25	10.075	0.683	0.415	343.7	67.1	4.41	116.5	23.1	2.57
	54	15.88	10.12	10.028	0.618	0.368	305.7	60.4	4.39	103.9	20.7	2.56
	49	14.40	10.00	10.000	0.558	0.340	272.9	54.6	4.35	93.0	18.6	2.54
10 × 8	45	13.24	10.12	8.022	0.618	0.350	248.6	49.1	4.33	53.2	13.3	2.00
	39	11.48	9.94	7.990	0.528	0.318	209.7	42.2	4.27	44.9	11.2	1.98
	33	9.71	9.75	7.964	0.433	0.292	170.9	35.0	4.20	36.5	9.2	1.94
10 × 5¾	29	8.53	10.22	5.799	0.500	0.289	157.3	30.8	4.29	15.2	5.2	1.34
	25	7.35	10.08	5.762	0.430	0.252	133.2	26.4	4.26	12.7	4.4	1.31
	21	6.19	9.90	5.750	0.340	0.240	106.3	21.5	4.14	9.70	3.4	1.25
8 × 8	67	19.70	9.00	8.287	0.933	0.575	271.8	60.4	3.71	88.6	21.4	2.12
	58	17.06	8.75	8.222	0.808	0.510	227.3	52.0	3.65	74.9	18.2	2.10
	48	14.11	8.50	8.117	0.683	0.405	183.7	43.2	3.61	60.9	15.0	2.08
	40	11.76	8.25	8.077	0.558	0.365	146.3	35.5	3.53	49.0	12.1	2.04
	35	10.30	8.12	8.027	0.493	0.315	125.5	31.1	3.50	42.5	10.6	2.03
	31	9.12	8.00	8.000	0.433	0.288	109.7	27.4	3.47	37.0	9.2	2.01
8 × 6½	28	8.23	8.06	6.540	0.463	0.285	97.8	24.3	3.45	21.6	6.6	1.62
	24	7.06	7.93	6.500	0.398	0.245	82.5	20.8	3.42	18.2	5.6	1.61
8 × 5¼	20	5.88	8.14	5.268	0.378	0.248	69.2	17.0	3.43	8.50	3.2	1.20
	17	5.00	8.00	5.250	0.308	0.230	56.4	14.1	3.36	6.72	2.6	1.16

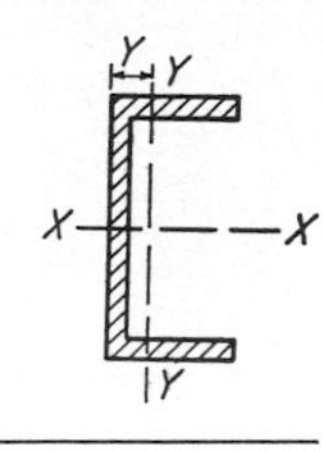

Properties of American Standard Channels

Nominal size	Depth of channel, in.	Weight per foot, lb	Area of section, in.²	Width of flange, in.	Web thickness, in.	Axis X-X			Axis Y-Y			
						I, in.⁴	S, in.³	r, in.	I, in.⁴	S, in.³	r, in.	y, in.
18 × 4	18	58.0	16.98	4.200	0.700	670.7	74.5	6.29	18.5	5.6	1.04	0.88
		51.9	15.18	4.100	0.600	622.1	69.1	6.40	17.1	5.3	1.06	0.87
		45.8	13.38	4.000	0.500	573.5	63.7	6.55	15.8	5.1	1.09	0.89
		42.7	12.48	3.950	0.450	549.2	61.0	6.64	15.0	4.0	1.10	0.90
15 × 3⅜	15	50.0	14.64	3.716	0.716	401.4	53.6	5.24	11.2	3.8	0.87	0.80
		40.0	11.70	3.520	0.520	346.3	46.2	5.44	9.3	3.4	0.89	0.78
		33.9	9.90	3.400	0.400	312.6	41.7	5.62	8.2	3.2	0.91	0.79
12 × 3	12	30.0	8.79	3.170	0.510	161.2	26.9	4.28	5.2	2.1	0.77	0.68
		25.0	7.32	3.047	0.387	143.5	23.9	4.43	4.5	1.9	0.79	0.68
		20.7	6.03	2.940	0.280	128.1	21.4	4.61	3.9	1.7	0.81	0.70
10 × 2⅝	10	30.0	8.80	3.033	0.673	103.0	20.6	3.42	4.0	1.7	0.67	0.65
		25.0	7.33	2.886	0.526	90.7	18.1	3.52	3.4	1.5	0.68	0.62
		20.0	5.86	2.739	0.379	78.5	15.7	3.66	2.8	1.3	0.70	0.61
		15.3	4.47	2.600	0.240	66.9	13.4	3.87	2.3	1.2	0.72	0.64
9 × 2½	9	20.0	5.86	2.648	0.448	60.6	13.5	3.22	2.4	1.2	0.65	0.59
		15.0	4.39	2.485	0.285	50.7	11.3	3.40	1.9	1.0	0.67	0.59
		13.4	3.89	2.430	0.230	47.3	10.5	3.49	1.8	0.97	0.67	0.61
8 × 2¼	8	18.75	5.49	2.527	0.487	43.7	10.9	2.82	2.0	1.0	0.60	0.57
		13.75	4.02	2.343	0.303	35.8	9.0	2.99	1.5	0.86	0.62	0.56
		11.5	3.36	2.260	0.220	32.3	8.1	3.10	1.3	0.79	0.63	0.58
7 × 2	7	14.75	4.32	2.299	0.419	27.1	7.7	2.51	1.4	0.79	0.57	0.53
		12.25	3.58	2.194	0.314	24.1	6.9	2.59	1.2	0.71	0.58	0.53
		9.8	2.85	2.090	0.210	21.1	6.0	2.72	0.98	0.63	0.59	0.55
6 × 2	6	13.0	3.81	2.157	0.437	17.3	5.8	2.13	1.1	0.65	0.53	0.52
		10.5	3.07	2.034	0.314	15.1	5.0	2.22	0.87	0.57	0.53	0.50
		8.2	2.39	1.920	0.200	13.0	4.3	2.34	0.70	0.50	0.54	0.52
5 × 1¾	5	9.0	2.63	1.885	0.325	8.8	3.5	1.83	0.64	0.45	0.49	0.48
		6.7	1.95	1.750	0.190	7.4	3.0	1.95	0.48	0.38	0.50	0.49
4 × 1⅝	4	7.25	2.12	1.720	0.320	4.5	2.3	1.47	0.44	0.35	0.46	0.46
		5.4	1.56	1.580	0.180	3.8	1.9	1.56	0.32	0.29	0.45	0.46
3 × 1½	3	6.0	1.75	1.596	0.356	2.1	1.4	1.08	0.31	0.27	0.42	0.46
		5.0	1.46	1.498	0.258	1.8	1.2	1.12	0.25	0.24	0.41	0.44
		4.1	1.19	1.410	0.170	1.6	1.1	1.17	0.20	0.21	0.41	0.44

Properties of Equal Angles

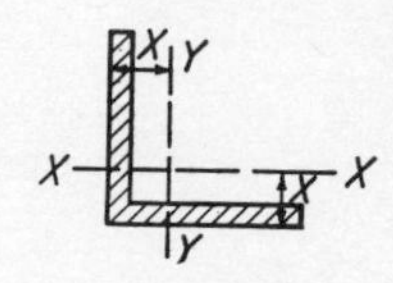

Size, in.	Thickness, in.	Weight per foot, lb	Area of section, in.²	Axis X-X and axis Y-Y			
				I, in.⁴	S, in.³	r, in.	x, in.
8 × 8	1⅛	56.9	16.73	98.0	17.5	2.42	2.41
	1	51.0	15.00	89.0	15.8	2.44	2.37
	⅞	45.0	13.23	79.6	14.0	2.45	2.32
	¾	38.9	11.44	69.7	12.2	2.47	2.28
	⅝	32.7	9.61	59.4	10.3	2.49	2.23
	9/16	29.6	8.68	54.1	9.3	2.50	2.21
	½	26.4	7.75	48.6	8.4	2.51	2.19
6 × 6	1	37.4	11.00	35.5	8.6	1.80	1.86
	⅞	33.1	9.73	31.9	7.6	1.81	1.82
	¾	28.7	8.44	28.2	6.7	1.83	1.78
	⅝	24.2	7.11	24.2	5.7	1.84	1.73
	9/16	21.9	6.43	22.1	5.1	1.85	1.71
	½	19.6	5.75	19.9	4.6	1.86	1.68
	7/16	17.2	5.06	17.7	4.1	1.87	1.66
	⅜	14.9	4.36	15.4	3.5	1.88	1.64
5 × 5	⅞	27.2	7.98	17.8	5.2	1.49	1.57
	¾	23.6	6.94	15.7	4.5	1.50	1.52
	⅝	20.0	5.86	13.6	3.9	1.52	1.48
	½	16.2	4.75	11.3	3.2	1.54	1.43
	7/16	14.3	4.18	10.0	2.8	1.55	1.41
	⅜	12.3	3.61	8.7	2.4	1.56	1.39
4 × 4	¾	18.5	5.44	7.7	2.8	1.19	1.27
	⅝	15.7	4.61	6.7	2.4	1.20	1.23
	½	12.8	3.75	5.6	2.0	1.22	1.18
	7/16	11.3	3.31	5.0	1.8	1.23	1.16
	⅜	9.8	2.86	4.4	1.5	1.23	1.14
	5/16	8.2	2.40	3.7	1.3	1.24	1.12
	¼	6.6	1.94	3.0	1.0	1.25	1.09
3½ × 3½	½	11.1	3.25	3.6	1.5	1.06	1.08
	7/16	9.8	2.87	3.3	1.3	1.07	1.04
	⅜	8.5	2.48	2.9	1.2	1.07	1.01
	5/16	7.2	2.09	2.5	0.98	1.08	0.99
	¼	5.8	1.69	2.0	0.79	1.09	0.97

PROPERTIES OF EQUAL ANGLES (*Continued*)

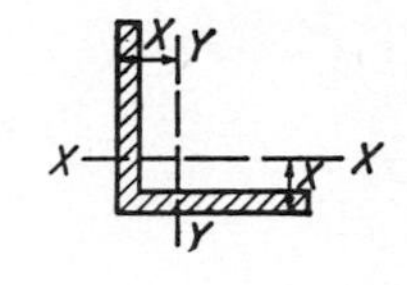

Size, in.	Thickness, in.	Weight per foot, lb	Area of section, in.²	Axis X-X and axis Y-Y			
				I, in.⁴	S, in.³	r, in.	x, in.
3 × 3	½	9.4	2.75	2.2	1.1	0.90	0.93
	7/16	8.3	2.43	2.0	0.95	0.91	0.91
	⅜	7.2	2.11	1.8	0.83	0.91	0.89
	5/16	6.1	1.78	1.5	0.71	0.92	0.87
	¼	4.9	1.44	1.2	0.58	0.93	0.84
2½ × 2½	½	7.7	2.25	1.2	0.73	0.74	0.81
	⅜	5.9	1.73	0.98	0.57	0.75	0.76
	5/16	5.0	1.47	0.85	0.48	0.76	0.74
	¼	4.1	1.19	0.70	0.39	0.77	0.72
	3/16	3.07	0.90	0.55	0.30	0.78	0.69
2 × 2	⅜	4.7	1.36	0.48	0.35	0.59	0.64
	5/16	3.92	1.15	0.42	0.30	0.60	0.61
	¼	3.19	0.94	0.35	0.25	0.61	0.59
	3/16	2.44	0.71	0.28	0.19	0.62	0.57
	⅛	1.65	0.48	0.19	0.13	0.63	0.55
1¾ × 1¾	¼	2.77	0.81	0.23	0.19	0.53	0.53
	3/16	2.12	0.62	0.18	0.14	0.54	0.51
	⅛	1.44	0.42	0.13	0.10	0.55	0.48
1½ × 1½	¼	2.34	0.69	0.14	0.13	0.45	0.47
	3/16	1.80	0.53	0.11	0.10	0.46	0.44
	⅛	1.23	0.36	0.08	0.07	0.46	0.42
1¼ × 1¼	¼	1.92	0.56	0.08	0.09	0.37	0.40
	3/16	1.48	0.43	0.06	0.07	0.38	0.38
	⅛	1.01	0.30	0.04	0.05	0.38	0.35
1 × 1	¼	1.49	0.44	0.06	0.06	0.29	0.34
	3/16	1.16	0.34	0.03	0.04	0.30	0.32
	⅛	0.80	0.23	0.02	0.03	0.31	0.30

Properties of Unequal Angles

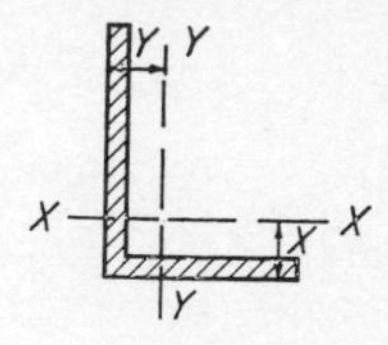

Size, in.	Thickness, in.	Weight per foot, lb	Area of section, in.^{2}	Axis X-X				Axis Y-Y			
				I, in.^{4}	S, in.^{3}	r, in.	x, in.	I, in.^{4}	S, in.^{3}	r, in.	y, in.
	1	44.2	13.00	80.8	15.1	2.49	2.65	38.8	8.9	1.73	1.65
	7/8	39.1	11.48	72.3	13.4	2.51	2.61	34.9	7.9	1.74	1.61
	3/4	33.8	9.94	63.4	11.7	2.53	2.56	30.7	6.9	1.76	1.56
8 × 6	5/8	28.5	8.36	54.1	9.9	2.54	2.52	26.3	5.9	1.77	1.52
	9/16	25.7	7.56	49.3	8.9	2.55	2.50	24.0	5.3	1.78	1.50
	1/2	23.0	6.75	44.3	8.0	2.56	2.47	21.7	4.8	1.79	1.47
	7/16	20.2	5.93	39.2	7.1	2.57	2.45	19.3	4.2	1.80	1.45
	1	37.4	11.00	69.6	14.1	2.52	3.05	11.6	3.9	1.03	1.05
	7/8	33.1	9.73	62.4	12.5	2.53	3.00	10.5	3.5	1.04	1.00
	3/4	28.7	8.44	54.9	10.9	2.55	2.95	9.4	3.1	1.05	0.95
8 × 4	5/8	24.2	7.11	46.9	9.2	2.56	2.91	8.1	2.6	1.07	0.91
	9/16	21.9	6.43	42.8	8.4	2.58	2.88	7.4	2.4	1.07	0.88
	1/2	19.6	5.75	38.5	7.5	2.59	2.86	6.7	2.2	1.08	0.86
	7/16	17.2	5.06	34.1	34.1	2.60	2.83	6.0	1.9	1.09	0.83
	1	34.0	10.00	47.7	10.8	2.18	2.60	11.2	3.9	1.06	1.10
	7/8	30.2	8.86	42.9	9.7	2.20	2.55	10.2	3.5	1.07	1.05
	3/4	26.2	7.69	37.8	8.4	2.22	2.51	9.1	3.0	1.09	1.01
	5/8	22.1	6.49	32.4	7.1	2.24	2.46	7.8	2.6	1.10	0.96
7 × 4	9/16	20.0	5.88	29.6	6.5	2.24	2.44	7.2	2.4	1.11	0.94
	1/2	17.9	5.25	26.7	5.8	2.25	2.42	6.5	2.1	1.11	0.92
	7/16	15.8	4.63	23.7	5.1	2.26	2.39	5.8	1.9	1.12	0.89
	3/8	13.6	3.99	20.6	4.4	2.27	2.37	5.1	1.6	1.13	0.87
	7/8	27.2	7.98	27.7	7.2	1.86	2.12	9.8	3.4	1.11	1.12
	3/4	23.6	6.94	24.5	6.2	1.88	2.08	8.7	3.0	1.12	1.08
	5/8	20.0	5.86	21.1	5.3	1.90	2.03	7.5	2.5	1.13	1.03
6 × 4	9/16	18.1	5.31	19.3	4.8	1.90	2.01	6.9	2.3	1.14	1.01
	1/2	16.2	4.75	17.4	4.3	1.91	1.99	6.3	2.1	1.15	0.99
	7/16	14.3	4.18	15.5	3.8	1.92	1.96	5.6	1.8	1.16	0.06
	3/8	12.3	3.61	13.5	3.3	1.93	1.94	4.9	1.6	1.17	0.94
	1/2	15.3	4.50	16.6	4.2	1.92	2.08	4.3	1.6	0.97	0.83
6 × 3½	7/16	13.5	3.97	14.8	3.7	1.93	2.06	3.8	1.4	0.98	0.81
	3/8	11.7	3.42	12.9	3.3	1.94	2.04	3.3	1.2	0.99	0.79
	3/4	19.8	5.81	13.9	4.3	1.55	1.75	5.6	2.2	0.98	1.00
	5/8	16.8	4.92	12.0	3.7	1.56	1.70	4.8	1.9	0.99	0.95
5 × 3½	1/2	13.6	4.00	10.0	3.0	1.58	1.66	4.0	1.6	1.01	0.91
	7/16	12.0	3.53	8.9	2.6	1.59	1.63	3.6	1.4	1.01	0.88
	3/8	10.4	3.05	7.8	2.3	1.60	1.61	3.2	1.2	1.02	0.86
	5/16	8.7	2.56	6.6	1.9	1.61	1.59	2.7	1.0	1.03	0.84

Properties of Unequal Angles (*Continued*)

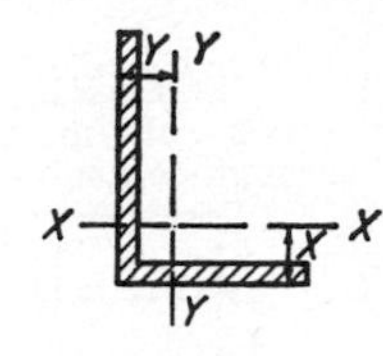

Size, in.	Thickness, in.	Weight per foot, lb	Area of section, in.^{2}	Axis X-X: I, in.^{4}	Axis X-X: S, in.^{3}	Axis X-X: r, in.	Axis X-X: x, in.	Axis Y-Y: I, in.^{4}	Axis Y-Y: S, in.^{3}	Axis Y-Y: r, in.	Axis Y-Y: y, in.
	5/8	14.7	4.30	6.4	2.4	1.22	1.29	4.5	1.8	1.03	1.04
	1/2	11.9	3.50	5.3	1.9	1.23	1.25	3.8	1.5	1.04	1.00
4 × 3½	7/16	10.6	3.09	4.8	1.7	1.24	1.23	3.4	1.3	1.05	0.98
	3/8	9.1	2.67	4.2	1.5	1.25	1.21	3.0	1.2	1.06	0.96
	5/16	7.7	2.25	3.6	1.3	1.26	1.18	2.6	1.0	1.07	0.93
	5/8	13.6	3.98	6.0	2.3	1.23	1.37	2.9	1.4	0.85	0.87
	1/2	11.1	3.25	5.0	1.9	1.25	1.33	2.4	1.1	0.86	0.83
4 × 3	7/16	9.8	2.87	4.5	1.7	1.25	1.30	2.2	1.0	0.87	0.80
	3/8	8.5	2.48	4.0	1.5	1.26	1.28	1.9	0.87	0.88	0.78
	5/16	7.2	2.09	3.4	1.2	1.27	1.26	1.7	0.74	0.89	0.76
	1/4	5.8	1.69	2.8	1.0	1.28	1.24	1.4	0.60	0.89	0.74
	1/2	10.2	3.00	3.5	1.5	1.07	1.13	2.3	1.1	0.88	0.88
	7/16	9.1	2.65	3.1	1.3	1.08	1.10	2.1	0.98	0.89	0.85
3½ × 3	3/8	7.9	2.30	2.7	1.1	1.09	1.08	1.8	0.85	0.90	0.83
	5/16	6.6	1.93	2.3	0.96	1.10	1.06	1.6	0.72	0.90	0.81
	1/4	5.4	1.56	1.9	0.78	1.11	1.04	1.3	0.58	0.91	0.79
	1/2	9.4	2.75	3.2	1.4	1.09	1.20	1.4	0.76	0.70	0.70
	7/16	8.3	2.43	2.9	1.3	1.09	1.18	1.2	0.68	0.71	0.68
3½ × 2½	3/8	7.2	2.11	2.6	1.1	1.10	1.16	1.1	0.59	0.72	0.66
	5/16	6.1	1.78	2.2	0.93	1.11	1.14	0.94	0.50	0.73	0.64
	1/4	4.9	1.44	1.8	0.75	1.12	1.11	0.78	0.41	0.74	0.61
	1/2	8.5	2.50	2.1	1.0	0.91	1.00	1.3	0.74	0.72	0.75
	7/16	7.6	2.21	1.9	0.93	0.92	0.98	1.2	0.66	0.73	0.73
3 × 2½	3/8	6.6	1.92	1.7	0.81	0.93	0.96	1.0	0.58	0.74	0.71
	5/16	5.6	1.62	1.4	0.69	0.94	0.93	0.90	0.49	0.74	0.68
	1/4	4.5	1.31	1.2	0.56	0.95	0.91	0.74	0.40	0.75	0.66
	1/2	7.7	2.25	1.9	1.0	0.92	1.08	0.67	0.47	0.55	0.58
	7/16	6.8	2.00	1.7	0.89	0.93	1.06	0.61	0.42	0.55	0.56
3 × 2	3/8	5.9	1.73	1.5	0.78	0.94	1.04	0.54	0.37	0.56	0.54
	5/16	5.0	1.47	1.3	0.66	0.95	1.02	0.47	0.32	0.57	0.52
	1/4	4.1	1.19	1.1	0.54	0.95	0.99	0.39	0.26	0.57	0.49
	3/8	5.3	1.55	0.91	0.55	0.77	0.83	0.51	0.36	0.58	0.58
2½ × 2	5/16	4.5	1.31	0.79	0.47	0.78	0.81	0.45	0.13	0.58	0.56
	1/4	3.62	1.06	0.65	0.38	0.78	0.79	0.37	0.25	0.59	0.54
	3/16	2.75	0.81	0.51	0.29	0.79	0.76	0.29	0.20	0.60	0.51
	1/4	2.77	0.81	0.32	0.24	0.62	0.66	0.15	0.14	0.43	0.41
2 × 1½	3/16	2.12	0.62	0.25	0.18	0.63	0.64	0.12	0.11	0.44	0.39
	1/8	1.44	0.42	0.17	0.13	0.64	0.62	0.09	0.08	0.45	0.37
	1/4	2.34	0.69	0.20	0.18	0.54	0.60	0.09	0.10	0.35	0.35
1¾ × 1¼	3/16	1.80	0.53	0.16	0.14	0.55	0.58	0.07	0.08	0.36	0.33
	1/8	1.23	0.36	0.11	0.09	0.56	0.56	0.05	0.05	0.37	0.31

Elements of Sections

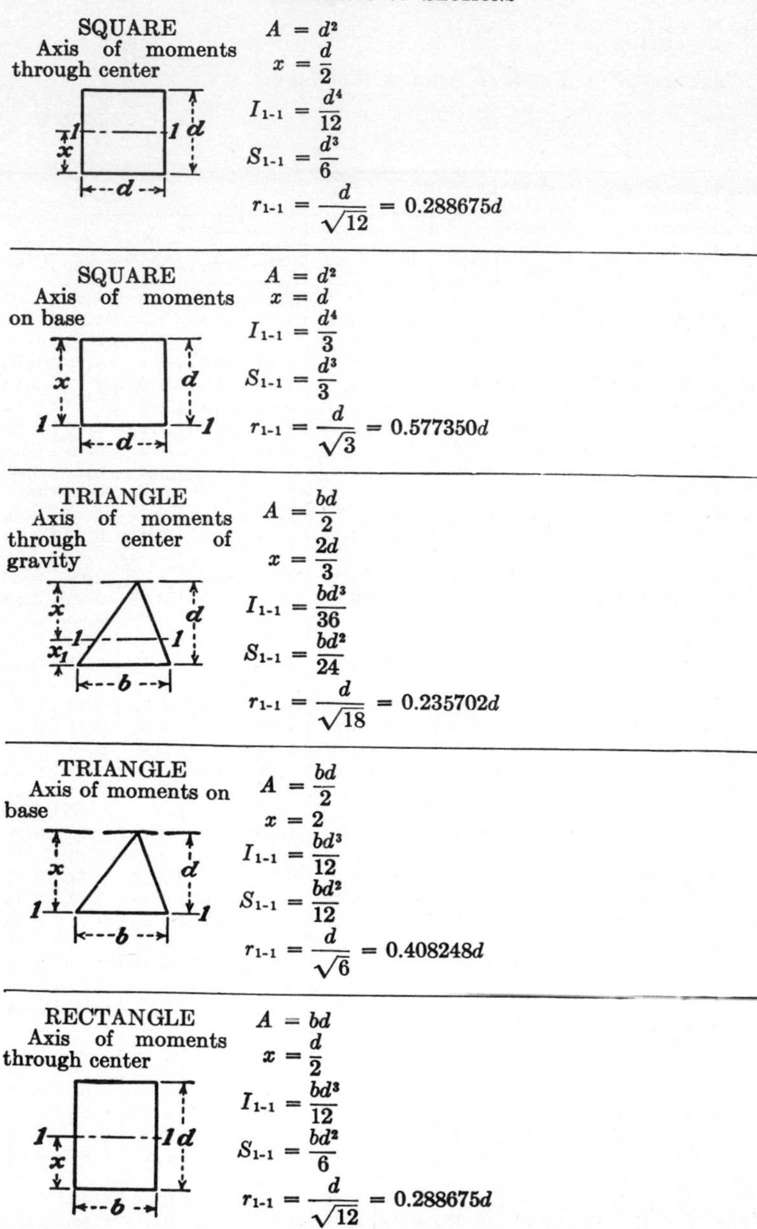

SQUARE
Axis of moments through center

$$A = d^2 \qquad x = \frac{d}{2} \qquad I_{1\text{-}1} = \frac{d^4}{12} \qquad S_{1\text{-}1} = \frac{d^3}{6} \qquad r_{1\text{-}1} = \frac{d}{\sqrt{12}} = 0.288675d$$

SQUARE
Axis of moments on base

$$A = d^2 \qquad x = d \qquad I_{1\text{-}1} = \frac{d^4}{3} \qquad S_{1\text{-}1} = \frac{d^3}{3} \qquad r_{1\text{-}1} = \frac{d}{\sqrt{3}} = 0.577350d$$

TRIANGLE
Axis of moments through center of gravity

$$A = \frac{bd}{2} \qquad x = \frac{2d}{3} \qquad I_{1\text{-}1} = \frac{bd^3}{36} \qquad S_{1\text{-}1} = \frac{bd^2}{24} \qquad r_{1\text{-}1} = \frac{d}{\sqrt{18}} = 0.235702d$$

TRIANGLE
Axis of moments on base

$$A = \frac{bd}{2} \qquad x = 2 \qquad I_{1\text{-}1} = \frac{bd^3}{12} \qquad S_{1\text{-}1} = \frac{bd^2}{12} \qquad r_{1\text{-}1} = \frac{d}{\sqrt{6}} = 0.408248d$$

RECTANGLE
Axis of moments through center

$$A = bd \qquad x = \frac{d}{2} \qquad I_{1\text{-}1} = \frac{bd^3}{12} \qquad S_{1\text{-}1} = \frac{bd^2}{6} \qquad r_{1\text{-}1} = \frac{d}{\sqrt{12}} = 0.288675d$$

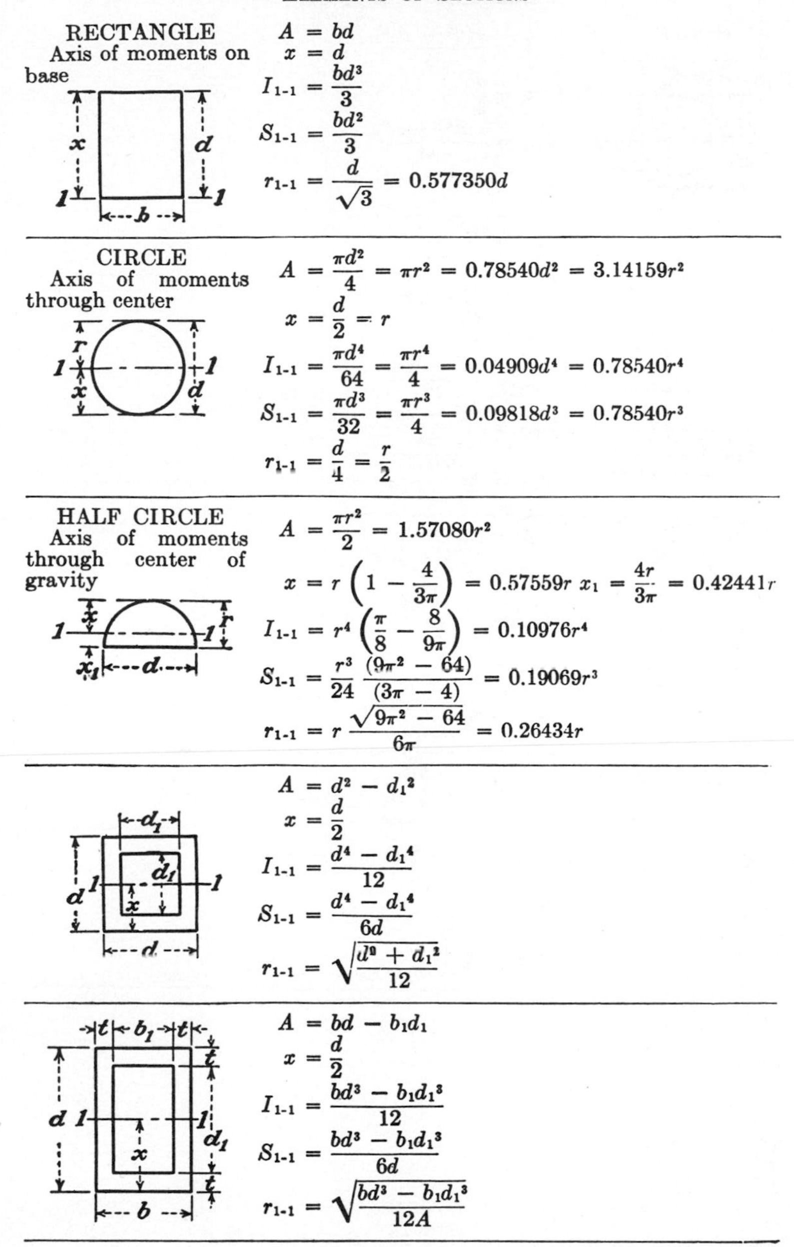

Elements of Sections

RECTANGLE
Axis of moments on base

$A = bd$

$x = d$

$I_{1-1} = \frac{bd^3}{3}$

$S_{1-1} = \frac{bd^2}{3}$

$r_{1-1} = \frac{d}{\sqrt{3}} = 0.577350d$

CIRCLE
Axis of moments through center

$A = \frac{\pi d^2}{4} = \pi r^2 = 0.78540d^2 = 3.14159r^2$

$x = \frac{d}{2} = r$

$I_{1-1} = \frac{\pi d^4}{64} = \frac{\pi r^4}{4} = 0.04909d^4 = 0.78540r^4$

$S_{1-1} = \frac{\pi d^3}{32} = \frac{\pi r^3}{4} = 0.09818d^3 = 0.78540r^3$

$r_{1-1} = \frac{d}{4} = \frac{r}{2}$

HALF CIRCLE
Axis of moments through center of gravity

$A = \frac{\pi r^2}{2} = 1.57080r^2$

$x = r\left(1 - \frac{4}{3\pi}\right) = 0.57559r \quad x_1 = \frac{4r}{3\pi} = 0.42441r$

$I_{1-1} = r^4\left(\frac{\pi}{8} - \frac{8}{9\pi}\right) = 0.10976r^4$

$S_{1-1} = \frac{r^3}{24}\frac{(9\pi^2 - 64)}{(3\pi - 4)} = 0.19069r^3$

$r_{1-1} = r\frac{\sqrt{9\pi^2 - 64}}{6\pi} = 0.26434r$

$A = d^2 - d_1^2$

$x = \frac{d}{2}$

$I_{1-1} = \frac{d^4 - d_1^4}{12}$

$S_{1-1} = \frac{d^4 - d_1^4}{6d}$

$r_{1-1} = \sqrt{\frac{d^2 + d_1^2}{12}}$

$A = bd - b_1d_1$

$x = \frac{d}{2}$

$I_{1-1} = \frac{bd^3 - b_1d_1^3}{12}$

$S_{1-1} = \frac{bd^3 - b_1d_1^3}{6d}$

$r_{1-1} = \sqrt{\frac{bd^3 - b_1d_1^3}{12A}}$

Elements of Sections

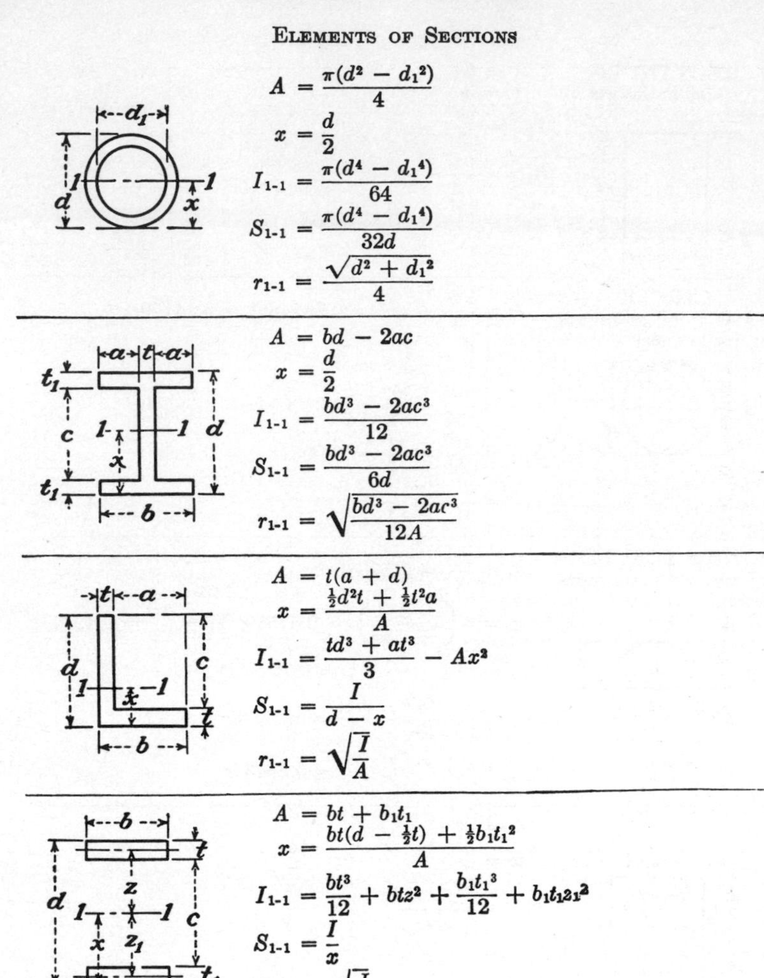

$$A = \frac{\pi(d^2 - d_1^2)}{4}$$
$$x = \frac{d}{2}$$
$$I_{1-1} = \frac{\pi(d^4 - d_1^4)}{64}$$
$$S_{1-1} = \frac{\pi(d^4 - d_1^4)}{32d}$$
$$r_{1-1} = \frac{\sqrt{d^2 + d_1^2}}{4}$$

$$A = bd - 2ac$$
$$x = \frac{d}{2}$$
$$I_{1-1} = \frac{bd^3 - 2ac^3}{12}$$
$$S_{1-1} = \frac{bd^3 - 2ac^3}{6d}$$
$$r_{1-1} = \sqrt{\frac{bd^3 - 2ac^3}{12A}}$$

$$A = t(a + d)$$
$$x = \frac{\frac{1}{2}d^2t + \frac{1}{2}t^2a}{A}$$
$$I_{1-1} = \frac{td^3 + at^3}{3} - Ax^2$$
$$S_{1-1} = \frac{I}{d - x}$$
$$r_{1-1} = \sqrt{\frac{I}{A}}$$

$$A = bt + b_1t_1$$
$$x = \frac{bt(d - \frac{1}{2}t) + \frac{1}{2}b_1t_1^2}{A}$$
$$I_{1-1} = \frac{bt^3}{12} + btz^2 + \frac{b_1t_1^3}{12} + b_1t_1z_1^2$$
$$S_{1-1} = \frac{I}{x}$$
$$r_{1-1} = \sqrt{\frac{I}{A}}$$

Weights and Areas of Bars

Size, inches	Weight, pounds per foot □	Weight, pounds per foot ○	Area, square inches □	Area, square inches ○
0				
1/16	.013	.010	.0039	.0031
1/8	.053	.042	.0156	.0123
3/16	.120	.094	.0352	.0276
1/4	.213	.167	.0625	.0491
5/16	.332	.261	.0977	.0767
3/8	.478	.376	.1406	.1105
7/16	.651	.511	.1914	.1503
1/2	.850	.668	.2500	.1963
9/16	1.076	.845	.3164	.2485
5/8	1.328	1.043	.3906	.3068
11/16	1.607	1.262	.4727	.3712
3/4	1.913	1.502	.5625	.4418
13/16	2.245	1.763	.6602	.5185
7/8	2.603	2.044	.7656	.6013
15/16	2.988	2.347	.8789	.6903
1	3.400	2.670	1.0000	.7854
1/16	3.838	3.015	1.1289	.8866
1/8	4.303	3.380	1.2656	.9940
3/16	4.795	3.766	1.4102	1.1075
1/4	5.313	4.172	1.5625	1.2272
5/16	5.857	4.600	1.7227	1.3530
3/8	6.428	5.049	1.8906	1.4849
7/16	7.026	5.518	2.0664	1.6230
1/2	7.650	6.008	2.2500	1.7671
9/16	8.301	6.519	2.4414	1.9175
5/8	8.978	7.051	2.6406	2.0739
11/16	9.682	7.604	2.8477	2.2365
3/4	10.413	8.178	3.0625	2.4053
13/16	11.170	8.773	3.2852	2.5802
7/8	11.953	9.388	3.5156	2.7612
15/16	12.763	10.024	3.7539	2.9483
2	13.600	10.681	4.0000	3.1416
1/16	14.463	11.359	4.2539	3.3410
1/8	15.353	12.058	4.5156	3.5466
3/16	16.270	12.778	4.7852	3.7583
1/4	17.213	13.519	5.0625	3.9761
5/16	18.182	14.280	5.3477	4.2000
3/8	19.178	15.062	5.6406	4.4301
7/16	20.201	15.866	5.9414	4.6664
1/2	21.250	16.690	6.2500	4.9087
9/16	22.326	17.534	6.5664	5.1572
5/8	23.428	18.400	6.8906	5.4119
11/16	24.557	19.287	7.2227	5.6727
3/4	25.713	20.195	7.5625	5.9396
13/16	26.895	21.123	7.9102	6.2126
7/8	28.103	22.072	8.2656	6.4918
15/16	29.338	23.042	8.6289	6.7771
3	30.60	24.03	9.000	7.069
1/16	31.89	25.05	9.379	7.366
1/8	33.20	26.08	9.766	7.670
3/16	34.54	27.13	10.160	7.980
3 1/4	35.91	28.21	10.563	8.296
5/16	37.31	29.30	10.973	8.618
3/8	38.73	30.42	11.391	8.946
7/16	40.18	31.55	11.816	9.281
1/2	41.65	32.71	12.250	9.621
9/16	43.15	33.89	12.691	9.968
5/8	44.68	35.09	13.141	10.321
11/16	46.23	36.31	13.598	10.680
3/4	47.81	37.55	14.063	11.045
13/16	49.42	38.81	14.535	11.416
7/8	51.05	40.10	15.016	11.793
15/16	52.71	41.40	15.504	12.177
4	54.40	42.73	16.000	12.566
1/16	56.11	44.07	16.504	12.962
1/8	57.85	45.44	17.016	13.364
3/16	59.62	46.83	17.535	13.772
1/4	61.41	48.23	18.063	14.186
5/16	63.23	49.66	18.598	14.607
3/8	65.08	51.11	19.141	15.033
7/16	66.95	52.58	19.691	15.466
1/2	68.85	54.07	20.250	15.904
9/16	70.78	55.59	20.816	16.349
5/8	72.73	57.12	21.391	16.800
11/16	74.71	58.67	21.973	17.257
3/4	76.71	60.25	22.563	17.721
13/16	78.74	61.85	23.160	18.190
7/8	80.80	63.46	23.766	18.665
15/16	82.89	65.10	24.379	19.147
5	85.00	66.76	25.000	19.635
1/16	87.14	68.44	25.629	20.129
1/8	89.30	70.14	26.266	20.629
3/16	91.49	71.86	26.910	21.135
1/4	93.71	73.60	27.563	21.648
5/16	95.96	75.36	28.223	22.166
3/8	98.23	77.15	28.891	22.691
7/16	100.53	78.95	29.566	23.221
1/2	102.85	80.78	30.250	23.758
9/16	105.20	82.62	30.941	24.301
5/8	107.58	84.49	31.641	24.850
11/16	109.98	86.38	32.348	25.406
3/4	112.41	88.29	33.063	25.967
13/16	114.87	90.22	33.785	26.535
7/8	117.35	92.17	34.516	27.109
15/16	119.86	94.14	35.254	26.688
6	122.40	96.13	36.000	28.274
1/16	124.96	98.15	36.754	28.866
1/8	127.55	100.18	37.516	29.465
3/16	130.17	102.23	38.285	30.069
1/4	132.81	104.31	39.063	30.680
5/16	135.48	106.41	39.848	31.296
3/8	138.18	108.53	40.641	31.919
7/16	140.90	110.66	41.441	32.548

625

Weights and Areas of Bars

Size, inches	Weight, pounds per foot		Area, square inches	
	□	○	□	○
6				
1/2	143.65	112.82	42.250	33.183
9/16	146.43	115.00	43.066	33.824
5/8	149.23	117.20	43.891	34.472
11/16	152.06	119.43	44.723	35.125
3/4	154.91	121.67	45.563	35.785
13/16	157.79	123.93	46.410	36.450
7/8	160.70	126.22	47.266	37.122
15/16	163.64	128.52	48.129	37.800
7	166.60	130.85	49.000	38.485
1/16	169.59	133.19	49.879	39.175
1/8	172.60	135.56	50.766	39.871
3/16	175.64	137.95	51.660	40.574
1/4	178.71	140.36	52.563	41.282
5/16	181.81	142.79	53.473	41.997
3/8	184.93	145.24	54.391	42.718
7/16	188.07	147.71	55.316	43.445
1/2	191.25	150.21	56.250	44.179
9/16	194.45	152.72	57.191	44.918
5/8	197.68	155.26	58.141	45.664
11/16	200.93	157.81	59.098	46.415
3/4	204.21	160.39	60.063	47.173
13/16	207.52	162.99	61.035	47.937
7/8	210.85	165.60	62.016	48.707
15/16	214.21	168.24	63.004	49.483
8	217.60	170.90	64.000	50.265
1/16	221.01	173.58	65.004	51.054
1/8	224.45	176.29	66.016	51.849
3/16	227.92	179.01	67.035	52.649
1/4	231.41	181.75	68.063	53.456
5/16	234.93	184.52	69.098	54.269
3/8	238.48	187.30	70.141	55.088
7/16	242.05	190.11	71.191	55.914
1/2	245.65	192.93	72.250	56.745
9/16	249.28	195.78	73.316	57.583
5/8	252.93	198.65	74.391	58.426
11/16	256.61	201.54	75.473	59.276
3/4	260.31	204.45	76.563	60.132
13/16	264.04	207.38	77.660	60.994
7/8	267.80	210.33	78.766	61.863
15/16	271.59	213.31	79.879	62.737
9	275.40	216.30	81.000	63.617
1/16	279.24	219.31	82.129	64.504
1/8	283.10	222.35	83.266	65.397
3/16	286.99	225.41	84.410	66.296

Size, inches	Weight, pounds per foot		Area, square inches	
	□	○	□	○
9				
1/4	290.91	228.48	85.563	67.201
5/16	294.86	231.58	86.723	68.112
3/8	298.83	234.70	87.891	69.029
7/16	302.83	237.84	89.066	69.953
1/2	306.85	241.00	90.250	70.882
9/16	310.90	244.18	91.441	71.818
5/8	314.98	247.38	92.641	72.760
11/16	319.08	250.61	93.848	73.708
3/4	323.21	253.85	95.063	74.662
13/16	327.37	257.12	96.285	75.622
7/8	331.55	260.40	97.516	76.589
15/16	335.76	263.71	98.754	77.561
10	340.00	267.04	100.000	78.540
1/16	344.26	270.38	101.254	79.525
1/8	348.55	273.75	102.516	80.516
3/16	352.87	277.14	103.785	81.513
1/4	357.21	280.55	105.063	82.516
5/16	361.58	283.99	106.348	83.525
3/8	365.98	287.44	107.641	84.541
7/16	370.40	290.91	108.941	85.563
1/2	374.85	294.41	110.250	86.590
9/16	379.33	297.92	111.566	87.624
5/8	383.83	301.46	112.891	88.664
11/16	388.36	305.02	114.223	89.710
3/4	392.91	308.59	115.563	90.763
13/16	397.49	312.19	116.910	91.821
7/8	402.10	315.81	118.266	92.886
15/16	406.74	319.45	119.629	93.957
11	411.40	323.11	121.000	95.033
1/16	416.09	326.80	122.379	96.116
1/8	420.80	330.50	123.766	97.205
3/16	425.54	334.22	125.160	98.301
1/4	430.31	337.97	126.563	99.402
5/16	435.11	341.73	127.973	100.510
3/8	439.93	345.52	129.391	101.623
7/16	444.78	349.33	130.816	102.743
1/2	449.65	353.16	132.250	103.869
9/16	454.55	357.00	133.691	105.001
5/8	459.48	360.87	135.141	106.139
11/16	464.43	364.76	136.598	107.284
3/4	469.41	368.68	138.063	108.434
13/16	474.42	372.61	139.535	109.591
7/8	479.45	376.56	141.016	110.754
15/16	484.51	380.54	142.504	111.923
12	489.60	384.53	144.000	113.098

Screw Threads and Bolts

American Standard Free Fit—Class 2

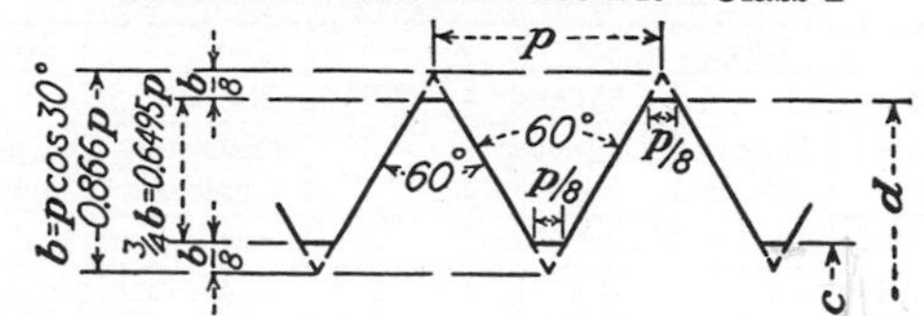

Diameter		Area			Diameter		Area		
Total d, inches	Net c, inches	Total diameter, d, square inches	Net diameter, c, square inches	Number of threads per inch	Total d, inches	Net c, inches	Total diameter, d, square inches	Net diameter, c, square inches	Number of threads per inch
¼	.185	.049	.027	20	2½	2.175	4.909	3.716	4
⅜	.294	.110	.068	16	2¾	2.425	5.940	4.619	4
½	.400	.196	.126	13					
⅝	.507	.307	.202	11	3	2.629	7.069	5.428	3¼
¾	.620	.442	.302	10	3¼	2.879	8.296	6.509	3½
⅞	.731	.601	.419	9	3½	3.100	9.621	7.549	3¼
					3¾	3.317	11.045	8.641	3
1	.838	.785	.551	8					
1⅛	.939	.994	.693	7	4	3.567	12.566	9.993	3
1¼	1.064	1.227	.890	7	4¼	3.798	14.186	11.330	2⅞
1⅜	1.158	1.485	1.054	6	4½	4.028	15.904	12.741	2¾
1½	1.283	1.767	1.294	6	4¾	4.255	17.721	14.221	2⅝
1⅝	1.389	2.074	1.515	5½					
1¾	1.490	2.405	1.744	5	5	4.480	19.635	15.706	2½
1⅞	1.615	2.761	2.049	5	5¼	4.730	21.648	17.574	2½
					5½	4.953	23.758	19.268	2⅜
2	1.711	3.142	2.300	4½	5¾	5.203	25.967	21.262	2⅜
2¼	1.961	3.976	3.021	4½					
					6	5.423	28.274	23.095	2¼

Bolt Heads and Nuts

United States and American Bridge Company Standard

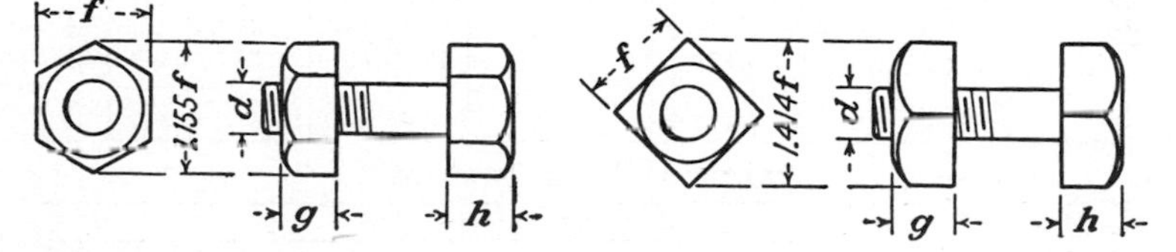

Heads and nuts		U. S. Standard	A. B. Co. Standard
Head	Height h	$0.75d + \frac{1}{16}$ in.	$0.75d$
	Short diameter f	$1.50d + \frac{1}{8}$ in.	$1.50d$
Nut	Height g	d	d
	Short diameter f	$1.50d + \frac{1}{8}$ in.	$1.50d + \frac{1}{8}$ in.

Heads for Bolts 1½ in. and under, A. B. Co. Standard.
Heads for Bolts 1⅝ in. and over, U. S. Standard.

Screw Threads and Bolts

American Bridge Company Standard

Diameter of bolt, inches	Head: Hexagon Diameter, inches, Long	Head: Hexagon Diameter, inches, Short	Head: Height, inches	Head: Square Diameter, inches, Long	Head: Square Diameter, inches, Short	Diameter of bolt, inches	Nut: Hexagon Diameter, inches, Long	Nut: Hexagon Diameter, inches, Short	Nut: Height, inches	Nut: Square Diameter, inches, Long	Nut: Square Diameter, inches, Short
1/4	7/16	3/8	3/16	1/2	3/8	1/4	9/16	1/2	1/4	11/16	1/2
3/8	5/8	9/16	1/4	3/4	9/16	3/8	11/16	11/16	3/8	1	11/16
1/2	7/8	3/4	3/8	1 1/16	3/4	1/2	1	7/8	1/2	1 1/4	7/8
5/8	1 1/8	15/16	1/2	1 3/8	15/16	5/8	1 1/4	1 1/16	5/8	1 1/2	1 1/16
3/4	1 1/4	1 1/8	9/16	1 5/8	1 1/8	3/4	1 1/2	1 1/4	3/4	1 3/4	1 1/4
7/8	1 1/2	1 5/16	5/8	1 7/8	1 5/16	7/8	1 5/8	1 7/16	7/8	2	1 7/16
1	1 3/4	1 1/2	3/4	2 1/8	1 1/2	1	1 7/8	1 5/8	1	2 1/4	1 5/8
1 1/8	2	1 11/16	7/8	2 3/8	1 11/16	1 1/8	2 1/8	1 13/16	1 1/8	2 5/8	1 13/16
1 1/4	2 1/8	1 7/8	15/16	2 5/8	1 7/8	1 1/4	2 1/4	2	1 1/4	2 7/8	2
1 3/8	2 3/8	2 1/16	1	2 7/8	2 1/16	1 3/8	2 1/2	2 3/16	1 3/8	3 1/8	2 3/16
1 1/2	2 5/8	2 1/4	1 1/8	3 1/8	2 1/4	1 1/2	2 3/4	2 3/8	1 1/2	3 3/8	2 3/8
1 5/8	3	2 9/16	1 1/4	3 5/8	2 9/16	1 5/8	3	2 9/16	1 5/8	3 5/8	2 9/16
1 3/4	3 1/8	2 3/4	1 3/8	3 7/8	2 3/4	1 3/4	3 1/8	2 3/4	1 3/4	3 7/8	2 3/4
1 7/8	3 3/8	2 15/16	1 1/2	4 1/8	2 15/16	1 7/8	3 3/8	2 15/16	1 7/8	4 1/8	2 15/16
2	3 5/8	3 1/8	1 9/16	4 3/8	3 1/8	2	3 5/8	3 1/8	2	4 3/8	3 1/8
2 1/4	4	3 1/2	1 3/4	5	3 1/2	2 1/4	4	3 1/2	2 1/4	5	3 1/2
2 1/2	4 1/2	3 7/8	1 15/16	5 1/2	3 7/8	2 1/2	4 1/2	3 7/8	2 1/2	5 1/2	3 7/8
2 3/4	4 7/8	4 1/4	2 1/8	6	4 1/4	2 3/4	4 7/8	4 1/4	2 3/4	6	4 1/4
3	5 3/8	4 5/8	2 5/16	6 1/2	4 5/8	3	5 3/8	4 5/8	3	6 1/2	4 5/8
3 1/4	5 3/4	5	2 1/2	7	5	3 1/4	5 3/4	5	3 1/4	7	5
3 1/2	6 1/4	5 3/8	2 11/16	7 5/8	5 3/8	3 1/2	6 1/4	5 3/8	3 1/2	7 5/8	5 3/8
3 3/4	6 5/8	5 3/4	2 7/8	8 1/8	5 3/4	3 3/4	6 5/8	5 3/4	3 3/4	8 1/8	5 3/4
4	7	6 1/8	3 1/16	8 5/8	6 1/8	4	7	6 1/8	4	8 5/8	6 1/8
4 1/4	7 1/2	6 1/2	3 1/4	9 1/4	6 1/2	4 1/4	7 1/2	6 1/2	4 1/4	9 1/4	6 1/2
4 1/2	8	6 7/8	3 7/16	9 3/4	6 7/8	4 1/2	8	6 7/8	4 1/2	9 3/4	6 7/8
4 3/4	8 3/8	7 1/4	3 5/8	10 1/4	7 1/4	4 3/4	8 3/8	7 1/4	4 3/4	10 1/4	7 1/4
5	8 3/4	7 5/8	3 13/16	10 3/4	7 5/8	5	8 3/4	7 5/8	5	10 3/4	7 5/8
5 1/4	9 1/4	8	4	11 1/4	8	5 1/4	9 1/4	8	5 1/4	11 1/4	8
5 1/2	9 5/8	8 3/8	4 3/16	11 7/8	8 3/8	5 1/2	9 5/8	8 3/8	5 1/2	11 7/8	8 3/8
5 3/4	10 1/8	8 3/4	4 3/8	12 3/8	8 3/4	5 3/4	10 1/8	8 3/4	5 3/4	12 3/8	8 3/4
6	10 1/2	9 1/8	4 9/16	12 7/8	9 1/8	6	10 1/2	9 1/8	6	12 7/8	9 1/8

Length of Bolt Threads

Length of bolt, inches	Diameter of bolt, inches: 1/4	3/8	1/2	5/8	3/4	7/8	1	1 1/8	1 1/4
1 to 1 1/2	3/4	3/4	1	1 1/4					
1 5/8 to 2	3/4	3/4	1	1 1/4	1 1/2	1 1/2			
2 1/8 to 2 1/2	3/4	3/4	1	1 1/4	1 1/2	1 3/4	1 3/4		
2 5/8 to 3	7/8	7/8	1	1 1/4	1 1/2	1 3/4	1 3/4	2 1/4	
3 1/8 to 4	7/8	7/8	1 1/4	1 1/4	1 1/2	1 3/4	1 3/4	2 1/4	2 1/2
4 1/8 to 8	1	1	1 1/4	1 1/2	1 3/4	2	2 1/4	2 1/2	2 3/4
8 1/8 to 12	1	1	1 1/2	1 3/4	2	2 1/4	2 1/2	3	3
12 1/8 to 20	1	1	1 1/2	2	2	2 1/4	2 1/2	3	3

Bolts are usually threaded about 3 times the diameter; in no case are standard bolts threaded closer to the head than 1/4 in.

COEFFICIENTS OF EXPANSION

Substance	Coefficient, *n*	
	Centigrade	Fahrenheit
Metals and alloys		
Aluminum, wrought	.0000231	.0000128
Brass	.0000188	.0000104
Bronze wire	.0000193	.0000107
Bronze	.0000181	.0000101
Copper	.0000168	.0000093
German silver	.0000183	.0000102
Gold	.0000150	.0000083
Iron, cast, gray	.0000106	.0000059
Iron, wrought	.0000120	.0000067
Iron wire	.0000124	.0000069
Lead	.0000286	.0000150
Nickel	.0000126	.0000070
Platinum	.0000090	.0000050
Platinum-iridium, 15% Ir	.0000081	.0000045
Silver	.0000192	.0000107
Steel, cast	.0000110	.0000061
Steel, hard	.0000132	.0000073
Steel, medium	.0000120	.0000067
Steel, soft	.0000110	.0000061
Tin	.0000210	.0000117
Zinc, rolled	.0000311	.0000173

ANSWERS TO SELECTED PROBLEMS

1-3. $W = 13{,}600$ lb.
1-5. $A = 3.77$ sq in.
1-7. $S = 11.3$ psi.
1-9. $A = 3.90$ sq in.
1-13. $e = 0.1408$ in.
1-15. $\tan B = 0.00225$.
1-17. F.S. $= 3.78$; $e = 0.0212$ in.
1-19. $e = 0.0222$ in.
1-21. Clearance $= 0.0667$ in.
1-23. $S = 4{,}040$ psi.
1-25. $\mu = 0.29$; $S = 17{,}250$ psi.
1-29. $U = 15.94$ in.-lb per cu in.
1-31. $S = 32{,}600$ psi.
1-33. $e = 98.7$ in.
1-35. $t = 65.6$°F.
1-37. $S = 16{,}080$ psi.

2-3. $P = 12{,}600$ lb.
2-5. $P = 17{,}700$ lb.
2-7. $P = 21{,}500$ lb.
2-11. $P = 72{,}200$ lb.
2-13. $n = 5$.
2-15. $n = 4$.
2-17. Eff. $= 87.5$ percent.
2-21. $S_t = 500$ psi.
2-23. $P = 16{,}200$ lb.
2-25. $S_s = 12{,}400$ psi; $S_b = 12{,}960$ psi; $S_t = 5{,}180$ psi.
2-27. $R = 176$ psi.
2-29. $L = 6.5$ in.
2-31. $D = 10.6$ in.; $E = 4.4$ in.
2-33. $P = 90{,}000$ lb.
2-35. $D = 4.74$ in.

3-1. $S_s = 15{,}300$ psi.
3-3. $D = 5.04$ in.
3-5. $\theta_F = 4.38°$ clockwise.
3-7. $L = 149.3$ in.
3-9. $S_s = 6{,}320$ psi.
3-13 $\theta_C = 0.0183$ rad.
3-15. $\theta_A = 2.76°$.
3-17. $S_s = 4{,}520$ psi.
3-19. $D = 4.58$ in.
3-21. $S_s = 6{,}920$ psi.
3-25. $T = 15{,}300$ in.-lb.
3-27. $S_s = 9{,}600$ psi.
3-29. $N = 167$.

4-1. $V_{\max} = \pm 6$ kips; $M_{\max} = +36$ ft-kips.
4-3. $V_{\max} = \pm 30$ kips; $M_{\max} = +112.5$ ft-kips.
4-5. $V_{\max} = \pm \mathrm{P}$; $M_{\max} = +\mathrm{PL}/4$.
4-7. $V_{\max} = -4\mathrm{P}$; $M_{\max} = +16\mathrm{P}$.
4-9. $V_{\max} = -1.5\mathrm{P}$; $M_{\max} = -\mathrm{PL}$.
4-11. $V_{\max} = \pm 10$ kips; $M_{\max} = -40$ ft-kips.
4-13. $V_{\max} = -108$ kips; $V = 0$ at 10.4 ft from R_L; $M_{\max} = +374.1$ ft-kips.
4-15. $V_{\max} = \pm P$; $M_{\max} = -3P$.
4-17. $V_{\max} = +3P$; $M_{\max} = -40P$.
4-19. $V_{\max} = \pm 81$ kips; $M_{\max} = +486$ ft-kips.
4-21. $V_{\max} = -2$ kips; $M_{\max} = \pm 12$ ft-kips.
4-25. $V_{\max} = +12$ kips; $M_{\max} = -36$ ft-kips.
4-27. $V_{\max} = +M/2L$; $M_{\max} = +M$.
4-30. $V_{\max} = -\frac{2}{3}P$; $M_{\max} = \pm PL/9$.

5-1. $b = 12$ in.
5-3. $I/c = 61.4$ in.^{3} Use 16 WF at 40 lb.
5-7. $S_t = +10{,}300$ psi; $S_c = -20{,}600$ psi.
5-9. $S_t = 5{,}310$ psi.
5-11. Required $I/c = 204$ in.^{3} Beam is safe.
5-13. $S = 16{,}700$ psi.

5-15. $S_s = 17{,}860$ psi. Beam is not safe.
5-17. $S = 1{,}200$ psi.
5-19. $S_s = 750$ psi; $S_s = 188$ psi.
5-21. $S = 7{,}640$ psi; $S = 3{,}820$ psi.
5-23. $S_s = 430$ psi.
5-25. $P = 15{,}670$ lb.
5-27. $I/c = 21.8$ in.^{3} Use 10 in. W_F at 25 lb.
5-29. $S_s = 5{,}940$ psi.

6-1. $\Delta = 0.667$ in.
6-3. $y = \dfrac{5PL^3}{48EI}$
6-5. $\rho = \dfrac{EI}{6(10)^5}$; $\rho = \dfrac{EI}{12(10)^5}$
6-7. $\Delta = \dfrac{5{,}446{,}000P}{EI}$
6-9. $\rho = 17{,}500$ in.
6-11. $P = 40{,}000$ lb.
6-13. Use 18 in. W_F at 70 lb.
6-15. $y_t = 0.4249$ in.
6-17. $b = 4.16$ in.
6-19. $y_{DC} = 0.0484$ in.
6-21. $y = 0.211$ in.
6-23. $y = 0.525$ in.
6-25. $U = 2.38$ in.-lb per cu in.
6-27. $\Delta_t = 0.378$ in.
6-29. $U_t = 2{,}520$ in.-lb.

7-1. $S_A = +3{,}000$ psi; $S_B = -1{,}000$ psi.
7-3. $S_{max} = +250$ psi; $S_{min} = -50$ psi.
7-9. $S_C = -23{,}000$ psi; $S_D = +25{,}000$ psi.
7-11. $P_2 = 4{,}000$ lb.
7-15. $P = 4{,}390$ lb.
7-17. $S_{max} = 11{,}200$ psi.
7-19. $S_{S_A} = 9{,}490$ psi
$S_{S_B} = 6{,}000$ psi
$S_{S_D} = 9{,}490$ psi
$S_{S_E} = 12{,}000$ psi
7-23. $S_{S_F} = 0$; $S_{S_G} = 13{,}400$ psi; $S_{S_H} = 12{,}000$ psi; $S_{S_I} = 13{,}400$ psi.
7-25. $S_{S_J} = 11{,}700$ psi; $S_{S_K} = 18{,}900$ psi; $S_{S_L} = 10{,}800$ psi.

8-1. $P = 26{,}600$ lb.
8-3. $P = 31{,}400$ lb.
8-5. $P = 138{,}800$ lb.
8-7. Use 8 in. W_F at 17 lb.
8-9. From tables, use 8×8 W_F at 48 lb.
8-13. $P = 66{,}500$
8-17. $P = 204{,}000$ lb.
8-19. $P = 4{,}720$ lb.
8-21. $P = 5{,}700$ lb.

INDEX

Allowable stress, 5
  in bending, 109, 110
  in riveted joints, pressure vessels, 42
    structural, 48
  in welded joints, 50
Angle sections, equal leg, 174
  unequal leg, 176
Angle of twist, 66
Areas of plane figures, 178
  bars, 181
  centroid of, 178
  moment of inertia, 178
  radius of gyration, 178
Axial loads, 2
  combined with bending, 134

Beam deflections, 119–129
Beams, 81–99
  assumptions in, 81
  bending of, 119
  bending moments in, 91
  cantilever, 82
  definition of, 81
  deflections of, 119–129
    table of, 124
  design of, 108–111
  elastic curve for, 119
Beams, elastic strain energy in, 128
  loads on, 83
    concentrated, 83
    nonuniform, 83
    uniform, 83
  moments in, 91
  neutral axis in, 104
  neutral plane in, 103
  overhanging, 82
  points of inflection, 94
  radius of curvature, 119
  reactions of, 83
  section modulus of, 108
  shear force in, 86
  simple, 82
  stresses in, 103–115
    assumptions of, 103
    bending, 103–108
    shearing, 111–115
  supports, 81–82
  types of, 82
  WF sections, 168
Bearing stresses in riveted joints, 42, 48
Bending, with axial loading, 134
  of beams, 119
  elastic strain energy in, 128

Bending moment, in beams, 91
  diagrams, 91
  sign of, 91
Bending stresses, 103–108
Biaxial strain, 14
Buckling of columns, 152

Circumferential stress, 46
Coefficient of expansion, 21
Columns, 151–163
  aluminum, 157
  buckling of, 152
  defined, 151
  slenderness ratio of, 153
  steel, 152
  timber, 157
Coupling, shaft, 71
    bolted, 72
    welded, 75
Curvature, radius of, 119

Deflection of beams, 119–129
  by superposition, 122
  table of, 124
Deformation, angular, 66
  axial, 10
  elastic, 11
Diagrams, bending moment, 91
  shear, 87
  stress-strain, 10

Eccentric loading of riveted connections, 142
Efficiency of riveted connections, 42
Elastic behavior, 11
Elastic curve, 94, 119
  deflection of, 119
Elastic curve, slope of, 119
Elastic limit, 11
Elastic strain energy (*see* Strain energy)
Elasticity, modulus of, 11

Factor of safety, 5
Fillet welds, 50

Hooke's law, 10
Horizontal shear stress, 112
Horsepower in shafts, 70

Inflection, point of, 94

Joints, allowable stresses in, structural, 48
    welded, 50
  riveted (*see* Riveted joints)
  welded, 49

Lateral strain, 13
Loads, biaxial, 140–142
  eccentric, 142–145
Longitudinal stress, 44

Moment diagrams, 91
Moments of inertia of areas, 178

Neutral axis, 103

Overhanging beam, 82

Pitch of rivets, 35
Plastic behavior, 11
Point of inflection, 94
Poisson's ratio, 12
Power, defined, 70
Pressure vessels, thin-walled, 43
Properties, of angles, 174, 176
  of channels, 173
  of WF sections, 168
Proportional limit, 11

Radius of curvature, 119
Ratio, Poisson's, 12
  slenderness, 153
Riveted joints, 34
  allowable stresses for, 42, 48
  biaxially loaded, 140
  eccentrically loaded, 142
  efficiency of, 42
  failures in, 36
  pitch, 35
  stresses in, 36
Riveting stresses, for pressure vessels, 42
  for structural joints, 48

Safety, factor of, 5
Screw threads, 183
Section modulus, 108
Shafts, 62
  couplings, bolted, 72
    welded, 75
  power transmission by, 70
Shear and connections, 31–55
Shear diagrams for beams, 87
Shear forces in beams, 86
Shear stress, in riveted joints, 36
  in shafts, 62
  in welded joints, 51
Shear stress, in WF sections, 114
Sign conventions, bending moments in beams, 91
  shear in beams, 87
  tension and compression, 15, 134
Slenderness ratio, 153
Slope of elastic curve, 119
  table of, 124
Strain, axial, 10
  defined, 1
  lateral, 13
  and stress, 1–24
  torsional, 66
Strain energy, axial, 17
  bending, 128
  biaxial, 19
  defined, 17
  torsional, 68
  triaxial, 19
Stress, allowable, 5, 42, 48, 50, 109, 111
  axial, 3
    with bending, 134–140
  bearing, 38
  bending, 104
  breaking, 12
  circumferential, 46
  compressive, 1
  defined, 1
  longitudinal, 44
  shear, 1, 31
    in beams, 111
    in riveted joints, 36
    in shafts, 62
    in WF beams, 114
  and strain, 1–24
  tensile (*see* Tensile stress)
  ultimate, 4
    table of, 4
Stress-strain behavior, 10–14

Stress-strain diagram, 11
Superposition, for beams, 122
  for biaxial stress, 19
  principle of, 14
  for triaxial stress, 19
Supports for beams, 81–82

Tables, bolts and threads, 183
  channels, 173
  coefficients of expansion, 185
  deflections of beams, 124
  elements of sections, 178
  equal angles, 174
  maximum moments, 96
  polar moments of inertia, 63
  properties of materials, 4
  round and square bars, 181
  timber, *k* factors, 160
    safe stresses, 158
  unequal angles, 176
Tables, WF sections, 168
Tensile stress, axial, 1
  in beams, 104
  circumferential, 46
  longitudinal, 44
  in riveted joints, 37
Thermal expansion, 21
Torque, defined, 61
Torsion, 61–76
  angle of twist, 66
  elastic strain energy, 68
  shear stress in, 62
  transmission of, 70

Ultimate strength, 4

Welded joints, 49
  allowable stresses in, 50
WF sections, properties of, 168